Proceedings of the 18th Symposium on Environmental Issues and Waste Management in Energy and Mineral Production

Eleonora Widzyk-Capehart
Asieh Hekmat · Raj Singhal
Editors

Proceedings of the 18th Symposium on Environmental Issues and Waste Management in Energy and Mineral Production

SWEMP 2018—Selected Works

Springer

Editors
Eleonora Widzyk-Capehart
AMTC, University of Chile
Santiago, Chile

Asieh Hekmat
DIMET, University of Concepción
Concepción, Chile

Raj Singhal
International Journal of Mining,
Reclamation and Environment
Calgary, AB, Canada

ISBN 978-3-319-99902-9 ISBN 978-3-319-99903-6 (eBook)
https://doi.org/10.1007/978-3-319-99903-6

Library of Congress Control Number: 2018955409

This Springer imprint is published by the registered company Springer Nature Switzerland AG
The registered company address is: Gewerbestrasse 11, 6330 Cham, Switzerland

Organizing Committee

International Chairs

Dr. Raj Singhal, Canada
Dr. Mohan Yellishetty, Australia

SWEMP 2018 Chairs

Dr. Eleonora Widzyk-Capehart, University of Chile
Dr. Asieh Hekmat, University of Concepcion

Technical Program Committee Chair

Dr. Andreina García

Co-Chairs

Dr. Petr Sklenicka, Czech Republic
Prof. Zhenqi Hu, China
Prof. Michael Hitch, Australia

International Organizing Committee

Prof. Sukumar Bandopadhyay, USA
Dr. Z. Bzowski, Poland
Prof. Carmen Mihaela Neculita, Canada
Prof. Josee Duchesne, Canada
Dr. Lidia Gawlik, Poland
Prof. Ge Hao, China
Prof. Gurdeep Singh, India
Prof. Giorgio Massacci, Italy
Dr. Maria Menegaki, Greece
Prof. Toyoharu Nawa, Japan
Dr. Antonio Nieto, USA
Dr. Bernadette O'Regan, Ireland
Prof. Kostas Fytas, Canada
Ms. M. Singhal, Canada
Prof. S. V. Yefremova, Kazakhstan
Prof. A. B. Szwilski, USA
Prof. Mauricio L. Torem, Brazil
Prof. Vladimir Kebo, Czech Republic
Prof. Morteza Osanloo, Iran
Dr. Behzad Ghodrati, Sweden
Dr. Valentina Dentoni, Italy
Prof. Takashi Sasaoka, Japan
Dr. Sanjay Kumar Shukla, Australia
Dr. Gento Mogi, Japan
Dr. Mohan Yellishetty, Australia
Ms. Yelena Yefremova, Germany
Prof. Semyon Shkundin, Russia
Dr. Nuray Demirel, Turkey
Prof. Wacław Dziurzyński, Poland
Prof. Stanisław Wasilewski, Poland
Dr. Jerzy Krawczyk, Poland
Dr. Mirosław Wierzbicki, Poland
Dr. Przemysław Skotniczny, Poland
Prof. Barbara Tora, Poland
Dr. Radosław Pomykała, Poland
Prof. Bekir Genc, South Africa
Dr. Steven Rupprecht, South Africa

Technical Review Committee

The Organizing Committee would like to express profound gratitude and appreciation to the following reviewers for their effort to improve the quality of the papers of this Proceeding.

Amir Adeli Sarcheshme, University of Chile, Chile
Ataç Başçetin, Istanbul University, Turkey
Shashibhushan Biliangadi, Monash University, Australia
Zbigniew Bzowski, Central Mining Institute, Poland
Valentina Dentoni, University of Cagliari, Italy
Wacław Dziurzyński, PAS Strata Mechanics Research Institute, Poland
Mansour Edraki, The University of Queensland, Australia
Andreina Garcia, University of Chile, AMTC, Chile
Bekir Genc, University of New South Wales, Australia
Asieh Hekmat, University of Concepcion, Chile
Michael Hitch, University of New South Wales, Australia
Saeed Karimi-Nasab, Shahid Bahonar University of Kerman, Iran
Giorgio Massacci, University of Cagliari, Italy
Maria Menegaki, National Technical University of Athens, Greece
Carmen Mihaela Neculita, University of Québec in Abitibi-Témiscamingue, Canada
Morteza Osanloo, AmirKabir University of Technology, Iran
Fhatuwani Sengani, University of the Witwatersrand, South Africa
Eleonora Widzyk-Capehart, University of Chile, AMTC, Chile
Mohan Yellishetty, Monash University, Australia
Tawanda Zvarivadza, University of the Witwatersrand, South Africa

Foreword

Raj Singhal

Mohan Yellishetty

This Symposium on Environmental Issues and Waste Management in Energy and Mineral Production (SWEMP) is the 18th in a series of biannual symposia on the subject matter. The basic aim of this series of symposia is to contribute to the development of methods and technologies for assessing, minimizing and preventing environmental problems connected with mineral and energy production.

This symposium has come to be recognized as a leader in promoting international technology transfer. A wide range of high-quality papers from North and South America, Europe, Australia, Africa and Asia have been attracted. Major topics to be covered are as follows: Control of Emissions in Mining Processes; Emerging Monitoring and Measurement Technologies; Environmental Chemistry and Toxicology: Health Hazard and Safety Issues; Environmental Issues in Deep Mining: Mine Ventilation, Refrigeration and Worker Health; Risk and Environmental Impact Assessment; Life Cycle Assessment; Management of Mining and Hazardous Waste and Waste Stabilization; Tailings Treatment, Recycle, Disposal; Mine Closure; Mine Rehabilitation and Reclamation; Remediation and Bioremediation; Water and Effluents: Treatment and Management; and Sustainability: Economic, Social and Climate Change.

SWEMP 2018 derives its strength from the coalition of various worldwide institutions. It is organized by the Advanced Mining Technology Center, University of Chile and University of Concepcion, Chile, in collaboration with the Department of Landscape Ecology, Faculty of Environmental Sciences, Czech University of Life Sciences, Czech Republic; Faculty of Agriculture University of South Bohemia, Czech Republic; The International Journal of Mining, Reclamation and Environment; The Department of Mining, Metallurgical and Materials Engineering, Universite Laval, Canada; National Technical University of Athens, Greece; University of Cagliari, Italy; Centre for Environmental Engineering Research and Education (CEERE), University of Calgary, Canada; Institute of Land Reclamation and Ecological Restoration, China University of Mining and Technology, Beijing, China; Columbia University, USA; Virginia Polytechnic Institute and State University, USA; Mining Engineering, McGill University; Imperial College London; Lulea University of Technology, Sweden; Faculty of Geoengineering, Mining and Geology, Wroclaw University of Technology, Poland; Hokkaido University, Mineral Resources Engineering Department, Japan; Polish Academy of Science, Poland, Finnish Environment Institute (SYKE), Finland; Mining Engineering Department, Istanbul University, Turkey; Edith Cowan University, Australia; Resources Engineering, Monash University, Australia; and others.

The organization and success of such a symposium are due mainly to the tireless efforts of many individuals, authors included. All members of the Organizing Committee and conference chairpersons have contributed greatly. The support of our plenary session speakers, invited speakers and co-chairs is gratefully acknowledged. In addition, recognition is accorded to my chairpersons of this symposium Dr. Asieh Hekmat, Dr. Eleonora Widzyk-Capehart and Dr. Andreina Garcia who together with their local Organizing Committee made SWEMP 2018 a success. I also wish to acknowledge the contribution of Mohini Singhal (my wife) who has been involved with SWEMP since its inception. She is a committee member of MPES/SWEMP organization and is an associate editor of the *International Journal of Mining, Reclamation and Environment*.

As the International Chair and Founder of this series of symposia, I would like to recognize the guidance and support of Rector, Prof. Ing. Petr Sklenicka, C.Sc., Czech Republic, our honorary chair. We are grateful to Dr. Patricio Aceituno, Dean, Facultad de Ciencias Fisicas y Matematicas, Universidad de Chile and Dr. Luis Moran T. Dean, Facultad de Ingenieria, Universidad de Concepcion for accepting to hold this symposium under their tutelage.

This symposium is designed to provide a forum for the presentation, discussion and debate of state-of-the-art and emerging technology in the field of environmental management. Authors from over 20 countries with backgrounds in science, technology and management representing government, industry and academia concerned with energy and mineral production have contributed to these Proceedings.

Calgary, Canada
Melbourne, Australia

Dr. Raj Singhal
Dr. Mohan Yellishetty
Chairs, International Organizing Committee

Preface

Eleonora Widzyk-Capehart Asieh Hekmat

During most of Chile's history, from 1500 to the present, mining has been an important economic activity: sixteenth-century mining was oriented towards the exploitation of gold placer deposits using encomienda labour; after a period of decline in the seventeenth century, mining resurged in the eighteenth and early nineteenth centuries, this time revolving chiefly around silver and, in the first half of the twentieth century, copper mining has come to the forefront.

Chile is a global mining power. It is the largest copper producer, supplying 32% of worldwide production and the third largest producer of molybdenum. It also occupies leading positions in the processing of other minerals. Mining has been a pillar of national progress for a long time. Mining has been a leading force to power development and attract investment; it is a crucial contributor to the progress made in the quality of life the country has enjoyed in recent decades. However, mining operations have the potential to affect the environment for future generations.

To be more environmentally sustainable, mining companies worldwide are making efforts to minimize the footprint of their activities throughout the mining cycle and after the completion of their activities. In Chile, the mining industry with the support of the Government and through the engagements with national and international scientific communities is taking significant steps towards sustainable mining by developing and integrating practices that reduce the environmental impact of mining operations. These practices include measures, such as reducing water and energy consumption, minimizing land disturbance and waste production, preventing soil, water and air pollution at mine sites, and conducting successful mine closure and reclamation activities.

In this context, the 18th International Symposium on Environmental Issues and Waste Management in Energy and Mineral Production (SWEMP 2018) is one of the most important events of the 2018, bringing together the scientists and the industry to share experiences and the latest advances towards innovative solutions in Santiago, Chile, in November 2018.

Contributions from SWEMP 2018 discuss methods and technologies for assessing, minimizing and preventing environmental problems associated with mineral and energy production. Topics include environmental impacts of harmful emissions and spontaneous combustion, risk and environmental impact assessments of mining, management of mining and hazardous waste, waste stabilization, tailings' treatment, stabilization and design, water and effluents treatment and management, mine ventilation, and emerging monitoring and measurement technologies.

We present you with the Proceedings of SWEMP 2018, which we hope would enable the holistic reflection and the practical application of "environmental issues and waste management" towards a sustainable future.

With Best Regards and Buena Suerte,

Santiago, Chile
Concepción, Chile

Dr. Eleonora Widzyk-Capehart
Dr. Asieh Hekmat
Chairs, SWEMP 2018 Organizing Committee

Acknowledgements

Because the History of our Mining Industry is written by all,
we thank Sierra Gorda SCM for their contribution to this important testimony.

Contents

Part I Emissions and Spontaneous Combustion

Comparative Evaluation of CO_2 Emissions in Europe and Turkey Using GIS 3
N. Demirel

Industrial Wind Erosion: PM Emission from the Erodible Flat Surfaces of Tailing Basins 15
V. Dentoni, B. Grosso, G. Massacci, M. Cigagna, C. Levanti, C. Corda and F. Pinna

Ash, Volatile Matter and Carbon Content Influence on Spontaneous Combustion Liability of Coal-Shales 29
M. Onifade and B. Genc

Effect of the Heat Input by Dolerite Intrusions and the Propensity for Spontaneous Combustion in the Highveld Coalfields, South Africa 39
E. R. Mokone, T. Zvarivadza and F. Sengani

Part II Risk and Environmental Impact Assessment

Mud Inflow Risk Assessment in Block Caving Operation Based on AHP Comprehensive Method 51
A. Hekmat, A. Anani, F. Tapia and I. Navia

The Impact of Sand Mining on the Fluvial Environment: Case Study of Nzhelele River in Limpopo Province, South Africa 67
F. Sengani and T. Zvarivadza

Evaluation of Factors Influencing Slope Instability: Case Study of the R523 Road Between Thathe Vondo and Khalvha Area in South Africa 81
F. Sengani and T. Zvarivadza

Environmental Issues with Best Management Practices in Energy and Mineral Production 91
Gurdeep Singh

Numerical Evaluation of Incremental Visual Impact 111
V. Dentoni, B. Grosso, G. Massacci, M. Cigagna and C. Levanti

Possible Environmental Risks Associated with Steel Slag: A Batch Study 121
S. Biliangadi, V. N. L. Wong, M. Yellishetty, A. Kumar Dikshit and S. Majumdar

Part III Management of Mining and Hazardous Waste and Waste Stabilization

Application of Fly Ash to Acidic Soil to Improve Plant Growth in Disturbed Land of Open-Cut Mining 129
A. Hamanaka, H. Yamasaki, T. Sasaoka, H. Shimada and S. Matsumoto

An Investigation of the Geotechnical Properties of Coal Combustion By-products from Matimba Power Station in Lephalale, South Africa 139
L. Magunde, F. Sengani and T. Zvarivadza

Environmental Management and Metal Recovery: Re-processing of Mining Waste at Montevecchio Site (SW Sardinia) 149
P. P. Manca, G. Massacci and C. Mercante

Part IV Mineral Processing and Tailings Treatment, Recycle, Disposal

Universal Flotation Reagent Produced from Plant Waste 163
S. Yefremova, L. Bunchuk, Yu. Sukharnikov, E. Li, A. Niyazov, S. Shalgimbayev, Yu. Hain and A. Zharmenov

Laboratory Studies to Examine the Effects of Adding Cement to Various Layers of a Surface Paste Tailings Storage 169
S. Tuylu, A. Bascetin and D. Adiguzel

Numerical Modelling of Pb-Zn Mine Tailing Dam Based on Soil Stability 181
A. Bascetin, S. Tuylu, D. Adiguzel, H. Eker and E. Odabas

In Situ Evaluation of Mechanical Properties of Phosphate Tailings for Exploring Reuse Potential: Case Study of a Phosphate Mine, South Africa 189
T. T. Mayisa, M. E. Nengovhela, F. Amponsah-Dacosta, F. Sengani and T. Zvarivadza

Part V Water and Effluents: Treatment and Management

Assessment of Groundwater Quality: Case Study of Tshivhasa, Limpopo Province, South Africa 205
F. Sengani and T. Zvarivadza

Coke-Based Carbon Adsorbent 217
S. Yefremova, A. Kablanbekov, K. Anarbekov, L. Bunchuk, A. Terlikbayeva and A. Zharmenov

Part VI Mine Ventilation

Analysis of Ventilation System and Assessment of Hazards in the Process of Progressing Liquidation of Workings in Mine 'S' 227
Wacław Dziurzyński, Marek Grzywacz and Jerzy Krawczyk

Part VII Emerging Monitoring and Measurement Technologies

Data Acquisition System for Position Tracking and Human-Selected Physiological and Environmental Parameters in Underground Mine 241
P. Stefaniak, J. Wodecki, A. Michalak, A. Wyłomańska and R. Zimroz

Development of a Dust Violation Control Tool from Plant Data 249
Mustafa Erkayaoglu

Finding the Right Time to Mine: A Real Options Analysis of Landfill Mining Projects 257
M. Menegaki, D. Damigos, A. Benardos and D. Kaliampakos

Part I
Emissions and Spontaneous Combustion

Comparative Evaluation of CO_2 Emissions in Europe and Turkey Using GIS

N. Demirel

1 Introduction

Turkey is a rapidly growing country with an increasing energy demand. Turkey's energy demand is supplied by domestic and import energy resources. Domestic energy resources include lignite, hard coal, oil, natural gas, hydroelectricity, geothermal, wood, animal and plant wastes, solar, and wind. The total lignite reserve is 11.5 billion tonnes which accounts for 7.7% of the world's lignite reserve, and the total hard coal reserve is 1.3 billion tonnes. Therefore, coal is the dominant energy source with significant reserve availability throughout the world and in Turkey as well. In the future, coal is expected to be a significant primary energy resource throughout the world (Tokgöz 2011). Figure 1 presents the share of coal in energy production among primary energy sources in Turkey (IEA 2016).

The growing environmental awareness and sensitivity related to global warming and climate change causes increasing public pressure on the utilization of fossil fuels especially coal because lignite shares the significant portion in climate change, which is induced mainly by CO_2 emissions and acidification impact categories, as seen in Fig. 2.

There have been several research studies conducted to investigate the CO_2 emission related to coal consumption for Turkey. Demirbaş and Bakış (2004) reviewed the status of Turkey's renewable energy resources and made future projections. The authors claimed that replacing more carbon-intensive fuels with renewables will contribute to the mitigation of urban pollution and CO_2 emissions. Kaygusuz (2004) reviewed the relationship between energy consumption and climate change mitigation in Turkey. This study reveals that hard coal and lignite will remain as a primary energy resource. Turkey's energy production using coal and

N. Demirel (✉)
Mining Engineering Department, Middle East Technical University, Ankara, Turkey
e-mail: ndemirel@metu.edu.tr

E. Widzyk-Capehart et al. (eds.), *Proceedings of the 18th Symposium on Environmental Issues and Waste Management in Energy and Mineral Production*,
https://doi.org/10.1007/978-3-319-99903-6_1

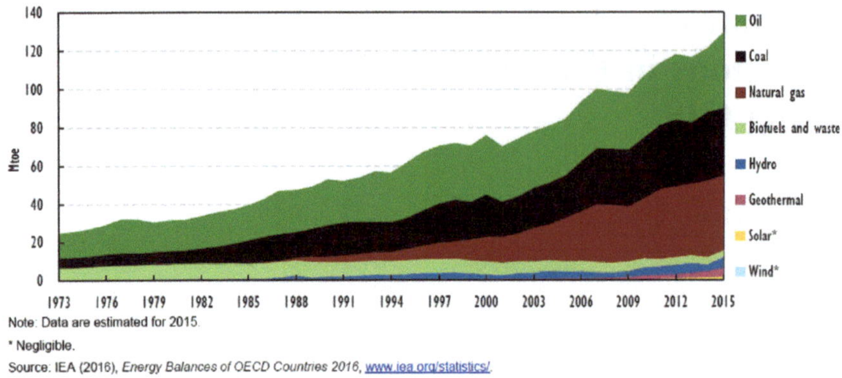

Fig. 1 Primary energy sources in Turkey (IEA 2016)

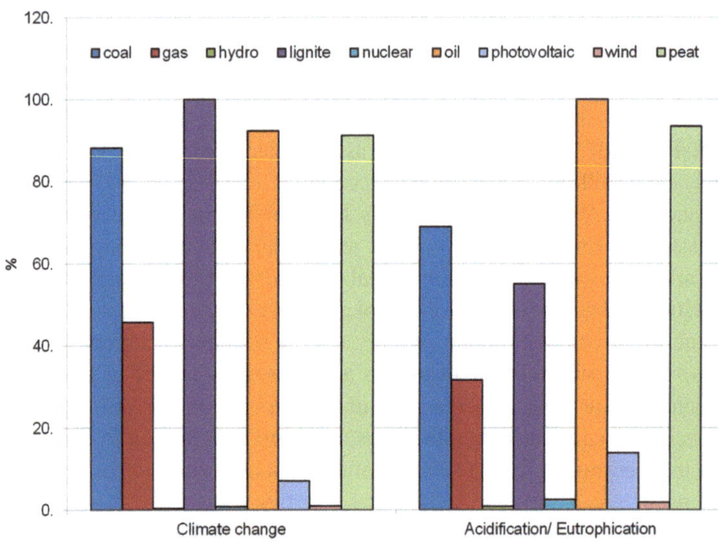

Fig. 2 Comparison of primary energy sources in terms of climate change and acidification impacts

lignite will be estimated to be 45,954 Kton in 2030, while it was 21,259 Kton in 2005 (MENR 2001). The author stated that encouraging and enforcing pollution control and environmental management measures for the mining sector is essential. Demirbaş (2006) reviewed Turkey's renewable energy resources and claimed that increasing use of domestic lignite has rapidly increased SO_2 emissions, which is mainly originated from the power sector, in recent years in Turkey (Demirbaş 2006). Say and Yücel (2006) conducted a study to overview the total energy consumption and CO_2 emissions in Turkey from 1970 to 2002. The authors performed a regression analysis between total energy consumption and total CO_2 emissions to be able to forecast the CO_2 emissions based on an economic growth and energy consumption for the future. It was claimed by the authors that the increase in the energy consumption would cause 9.9% increase in the CO_2 emission in average annually. Total CO_2 emission was estimated to increase from 480,244 Gg for 2010 to 631,781 Gg for 2015 using IPCC method. Yüksel and Sandalcı (2011) presented a review for the development of climate change, energy and environment in Turkey. The authors asserted that carbon intensity in Turkey is higher than the Western developed national average and the greenhouse gas emissions should be monitored and investigated regularly for policy development. Tokgöz (2011) developed a model to numerically evaluate the impact of CO_2 from fossil fuel consumption on global warming and climate change on a global scale. The author appealed that Turkey was affected by the CO_2 production of countries to the North and North-west and the CO_2 emission of countries in the West. It was stated that every year, 20.47 Giga tonnes of CO_2 was released into the atmosphere in the Northern Hemisphere coming over Turkey via atmospheric air movements from the West towards the East. Although all these research studies provided significant insight into Turkey's CO_2 emissions and energy consumption, none of them included a GIS-based comparative evaluation of coal consumption on climate change in Europe and Turkey.

This paper investigates the CO_2 emission trends of European countries with respect to their coal consumption and production for 26-year period from 1980 to 2006 using GIS. The research methodology followed in this study essentially entails four main stages: (i) collection and pre-processing of temporal-spatial data, (ii) generation of a database, (iii) performing GIS cluster analysis including cluster and outlier analysis and hot spot analysis, and (iv) interpreting the results.

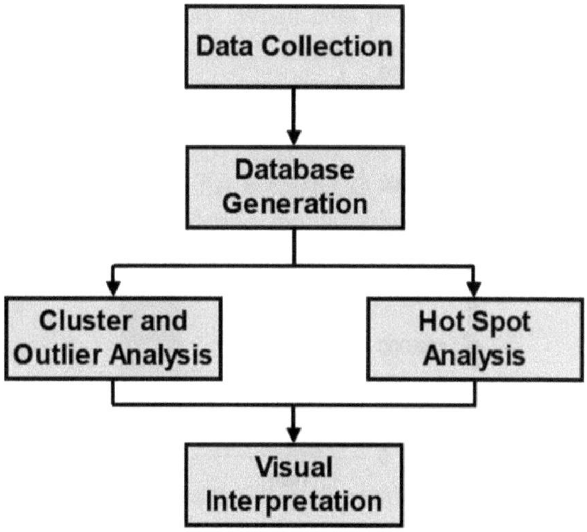

Fig. 3 Research framework followed in this study

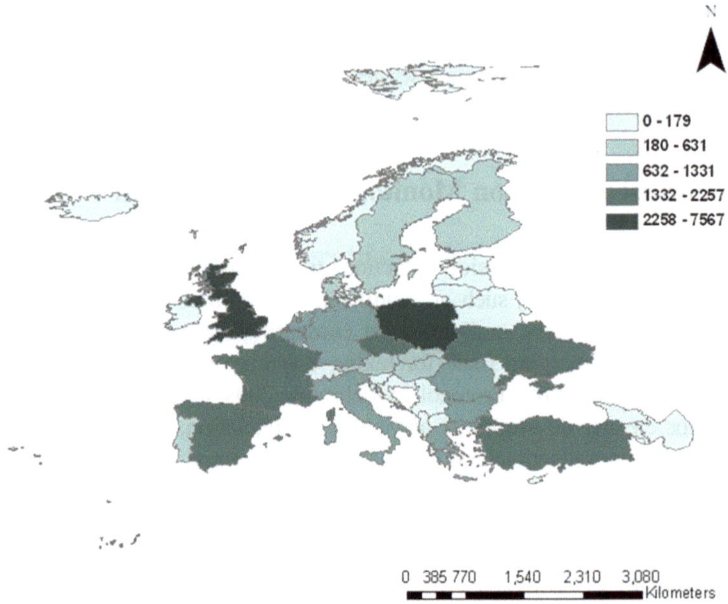

(a) CO_2 emission levels throughout Europe

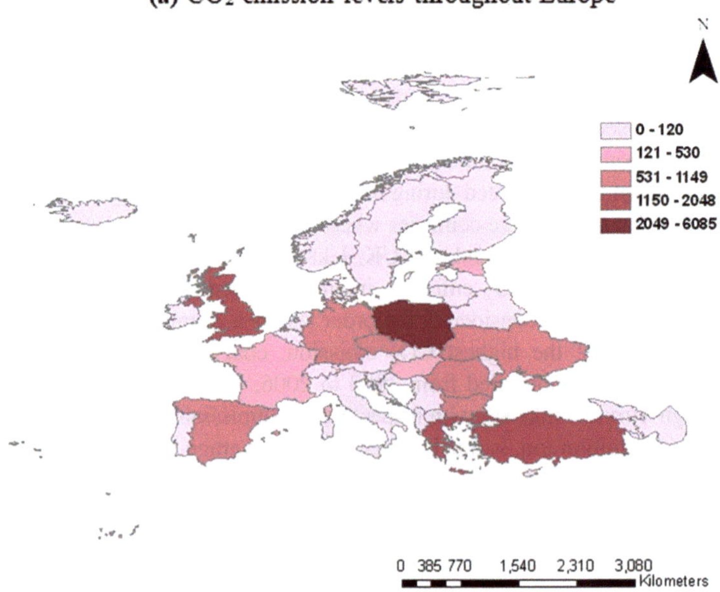

(b) Coal production levels throughout Europe

Fig. 5 CO_2 (**a**), coal production (**b**) and consumption (**c**) map for Europe

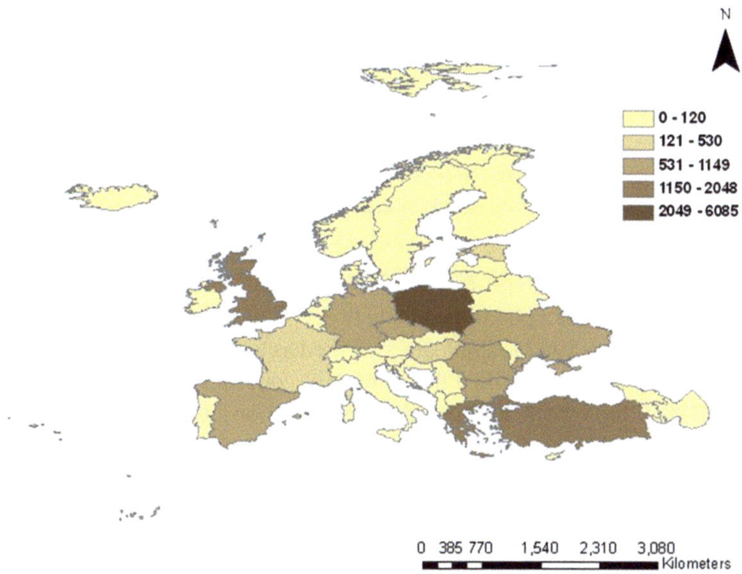

(c) Coal consumption levels throughout Europe

Fig. 5 (continued)

Table 1 First ten European countries in terms of CO_2 emissions, coal production, and consumption values

	CO_2 emissions (Mtons)		Coal production (Mtons)		Coal consumption (Mtons)	
1	Germany	11,250	Germany	10,228	Germany	10,856
2	Poland	7567	Poland	6085	Poland	5239
3	UK	5442	UK	2048	UK	2581
4	Ukraine	2257	Greece	1548	Turkey	1663
5	Spain	2233	Turkey	1430	Greece	1582
6	France	1831	Romania	1149	Spain	1441
7	Czech Republic	1790	Ukraine	1144	Romania	1326
8	Turkey	1656	Czech Republic	1061	Ukraine	1225
9	Italy	1331	Spain	1032	Bulgaria	1018
10	Romania	1196	Bulgaria	869	Czech Republic	985

feature values are also determined. Cluster and outlier analysis utilizes a Local Moran's I value, a *Z*-score, a *p*-value and a code representing the cluster type for each feature. The *Z*-score and *p*-value represent the statistical significance of the computed index value. Local Moran's I value is given in Eq. 1 (Mitchell 2005).

$$I_i = \frac{x_i - \bar{X}}{S_i^2} \sum_{j=1, j \neq 1}^{n} w_{i,j}(x_i - \bar{X}) \quad (1)$$

In Eq. 1, x_i is an attribute for feature i, $\overline{X}$ is the mean of the corresponding attribute, and $w_{i,j}$ is the spatial weight between feature and j. In this study, CO_2 emission values from 1980 to 2006 are identified as a set of values for the European countries. This analysis identified the country or countries having very low or high CO_2 emissions throughout Europe. Conceptualization of spatial relationship was chosen to be an inverse distance to specify how spatial relationships among features are conceptualized. In the inverse distance, all features impact all other features, but the farther away something is, the smaller the impact it has. Distance method, which specifies how distances are calculated when measuring spatial autocorrelation, was chosen as Euclidean distance. As it can be seen in Fig. 6, Poland and Germany were determined as the clusters differing from neighbouring countries with high CO_2 emission values.

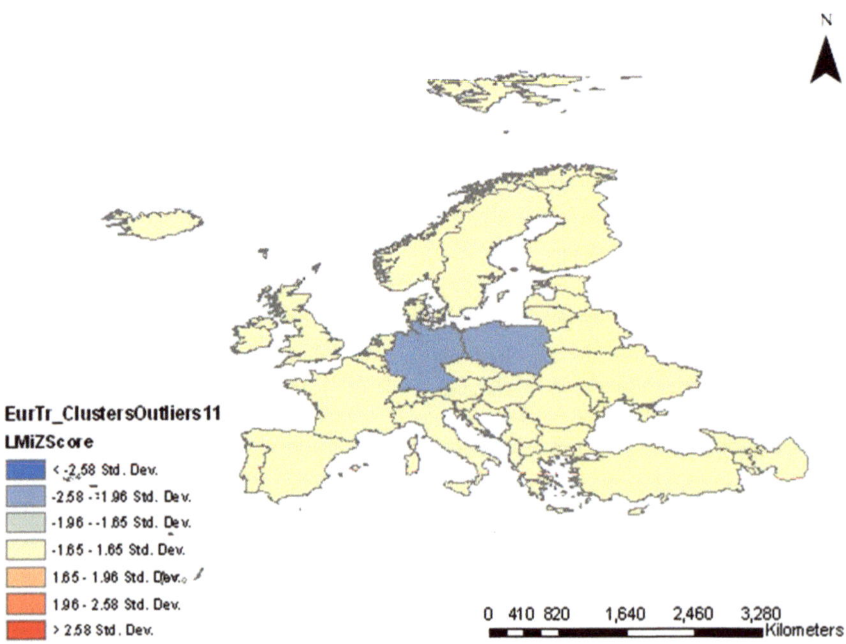

Fig. 6 Cluster and Outlier analysis results for the total CO_2 emission from 1980 to 2006

3.2 Hot Spot Analysis

In this study, the countries, which have high emission values and also are surrounded by countries having high emission values, are determined using hot spot analysis in GIS environment. In the hot spot analysis, the Getis-Ord Gi statistic for each feature in a data set is calculated and the spatial clusters of features with high or low values are determined using the resultant Z-score. This tool works by looking at each feature within the context of neighbouring features. A feature with a high value is interesting, but may not be a statistically significant hot spot. Getis-Ord Gi local statistic is given in Eq. 2 (Mitchell 2005).

$$G_i^* = \frac{\sum_{j=1}^{n} w_{i,j} x_j - \bar{X} \sum_{j=1}^{n} w_{i,j}}{S \sqrt{\frac{\left[n \sum_{j=1}^{n} w_{i,j}^2 - \left(\sum_{j=1}^{n} w_{i,j} \right)^2 \right]}{n-1}}} \quad (2)$$

In Eq. 2, x_j is the attribute value for feature j, $w_{i,j}$ is the spatial weight between feature i and j, and n is equal to the total number of features and:

$$\overline{X} = \frac{\sum_{j=1}^{n} x_j}{n} \quad (3)$$

$$S = \sqrt{\frac{\sum_{j=1}^{n} x_J^2}{n} - (\overline{X})^2} \quad (4)$$

To be statistically significant hot spot, a feature will have a high value and be surrounded by other features with high values as well. The local sum for a feature and its neighbours is compared proportionally to the sum of all features; when the local sum is more different than the expected local sum, that difference is too large to be the result of random chance, a statistically significant Z-score result. Figure 7 illustrates the results of the hot spot analysis. As it can be seen from Fig. 7, Belgium, Luxemburg, and Greece were determined to be the hot spots. Turkey was not determined to be a hot spot with its average score.

These two analyses, considering the total CO_2 emissions of European countries from 1980 to 2006, resulted that Turkey with its total CO_2 emission is not critical when compared to other European countries. In addition to these analysis, the map of CO_2 normalized with coal production was also generated (Fig. 8). Figure 8 illustrates that Turkey is one of the countries having the lowest normalized CO_2 with respect to coal production value of 0.5158–1.146. This means that, when CO_2 emissions normalized with the coal production is considered, Turkey can be regarded as a country influencing Europe's CO_2 emissions the least.

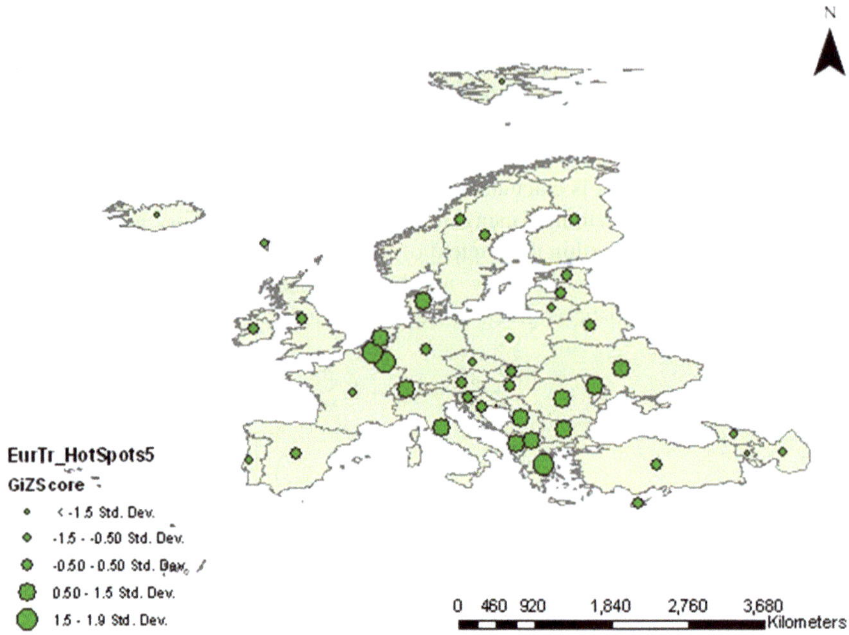

Fig. 7 Cluster CO_2 hot spot analysis

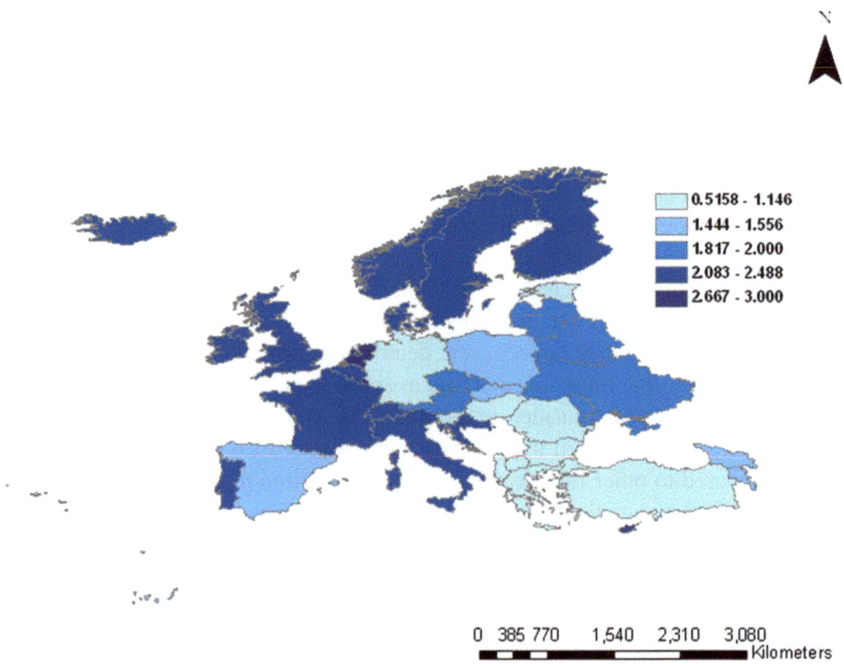

Fig. 8 Normalized CO_2 emission to coal production

4 Conclusions

In this study, comparative evaluation of CO_2 emissions of Turkey and European countries was carried out using GIS. Cluster and outlier analysis and hot spot analysis results revealed that Turkey, with total 1.656 million ton CO_2 emissions from 1980 to 2006, ranks 8th in Europe and Turkey's CO_2 emission is much lower than the average CO_2 emission values of the first ten countries which is 3,655 million ton. The normalized CO_2 emissions with coal production resulted that Turkey is one of the countries having the lowest normalized CO_2 with respect to coal production value of 0.5158–1.146. However, it is still essential to meet the guidelines and regulations to contribute the emission reduction efforts of Europe.

References

Demirbaş, A.: Turkey's renewable energy policy. Energy Sources Part A Recovery Utilization Environ Effects **28**, 657–665 (2006)

Demirbaş, A., Bakış, R.: Energy from renewable sources in Turkey: status and future direction. Energy Sources Part A Recovery Utilization Environ Effects **26**, 473–484 (2004)

IEA: Energy balances of OECD countries. www.iea.org/statistics/ (2016)

Kaygusuz, K.: Climate change mitigation in Turkey. Energy Sources Part A Recovery Utilization Environ Effects **26**, 563–573 (2004)

Ministry of Energy and Natural Resources (MENR): Energy report of Ankara, Turkey. http://www.menr.gov.trv (2001)

Mitchell, A.: The ESRI Guide to GIS Analysis, vol. 2. ESRI Press (2005)

Say, N.P., Yücel, M.: Energy consumption and CO_2 emissions in Turkey: empirical analysis and future projection based on an economic growth. Energy Policy **34**, 3870–3876 (2006)

Tokgöz, N.: A numerical analysis of worldwide CO_2 emissions based on fossil fuels and effects on atmospheric warming in Turkey. www.worldenergy.org/documents/p001555.doc. Last accessed 07 July 2011

Yüksel, I., Sandalcı, M.: Climate change, energy and the environment in Turkey. Energy Sources Part A Recovery Utilization Environ Effects **33**, 410–422 (2011)

Industrial Wind Erosion: PM Emission from the Erodible Flat Surfaces of Tailing Basins

V. Dentoni, B. Grosso, G. Massacci, M. Cigagna, C. Levanti, C. Corda and F. Pinna

1 Introduction

The emission of particulate matter (PM) from industrial sites typically derives from both conveyed sources (chimneys, dust collectors, etc.) and fugitive dust sources (material handling and transportation, heap formation, transit of vehicles along unpaved roads, wind erosion, etc.). While the emission from conveyed sources is relatively easy to estimate, the characterization of fugitive sources requires the knowledge of the physical properties of the handled/deposited material, the transportation cycle, and the type of machinery in use, as well as the anemological conditions of the site under consideration.

The dust flow deriving from fugitive sources is generally calculated as the product of the *action intensity* for a specific parameter (*Emission Factor*), which takes into account the source physical characteristics. The dust flow (kg/h) generated by handling operations of granular material is calculated, for example, as a product of the mass of material moved in the time unit (*action intensity*) by an *Emission Factor* (EF) that indicates the kilograms of dust emitted for each kilogram of material moved. The dust flow (kg/h) generated by earth moving vehicles travelling along unpaved roads (kg/h) is calculated as the product of the road length travelled in the time unit (*action intensity*) by an emission factor (EF) that indicates the kilograms of dust emitted by a road length unit (kg/km).

Dust emission caused by wind erosion is not linked to specific industrial operations but only to the wind action over the exposed surfaces of the material

V. Dentoni (✉) · B. Grosso · G. Massacci · C. Corda · F. Pinna
DICAAR—Department of Civil and Environmental Engineering and Architecture, University of Cagliari, Via Marengo 2, Cagliari, Italy
e-mail: vdentoni@unica.it

M. Cigagna · C. Levanti
CINIGEO—Consorzio Interuniversitario Nazionale per l'Ingegneria delle Georisorse, Rome, Italy

E. Widzyk-Capehart et al. (eds.), *Proceedings of the 18th Symposium on Environmental Issues and Waste Management in Energy and Mineral Production*,
https://doi.org/10.1007/978-3-319-99903-6_2

accumulated outdoor (heaps, dumps, tailing basins, etc.). The effect of the wind action depends on factors such as the extent and orientation of the deposit surfaces, the grain size and moisture content of the deposited material, and the anemological conditions of the specific site under exam. From this point of view, dust emission from tailing basins could be quite significant, due to both the extent of the basin surfaces exposed to wind and the small particle size of the disposed material.

The present article specifically deals with the emission of PM from the deposits of mineralogical processing residue. In fact, the examination of the technical and scientific reports has shown that the emission factors proposed for other types of erodible surfaces cannot be directly applied to those deposits, because of their peculiar characteristics: wide and flat surfaces with low roughness and residue physical state dependent on its moisture content.

The object of the research hereby discussed is the definition of an emission conceptual model applicable to the bauxite residue disposal areas (BRDA). In particular, based on the analysis of the scientific literature regarding wind erosion, the article proposes a specific-site conceptual model and its validation procedure.

2 Wind Erosion

The evaluation of the dust flow generated by wind erosion is particularly complex and is typically based on the development of specific-site conceptual models. The parameters that influence the erosion phenomenon are, in fact, numerous and of complex evaluation, so that general emission models only interpret the main laws governing the phenomenon, while the most complex and detailed aspects are taken into account by using constant parameters, whose values are decisive for each specific case study.

The comprehension of the wind erosion mechanism is of primary importance in the studies of landscape dynamics (formation or erosion of dunes, beaches, etc.), when analyzing problems of soil impoverishment in agricultural areas or assessing the environmental impact arising from industrial activities. In all cases mentioned above, the erosion phenomenon causes the emission and dispersion of granular materials composed of free inorganic particles. In the field of geological sciences, particles between 60 and 2000 μm moving in contact with the ground are of primary interest. Environmental and health impact studies consider smaller particles (PM_{10} and $PM_{2.5}$), as they are transported in suspension by the air and, most important, they might be capable of penetrating the inner parts of the human respiratory system.

The lifting mechanism of a solid particle is governed by the wind speed and by the particle size and density. The lifting action is explained by the drag force (F_d) and lifting force (F_L) exerted by the wind, which are opposed by the gravitational force (F_g) and by the surface adhesion force (F_{ip}). Shao and Lu (2000) described the motion trigger mechanism with reference to the scheme in Fig. 1.

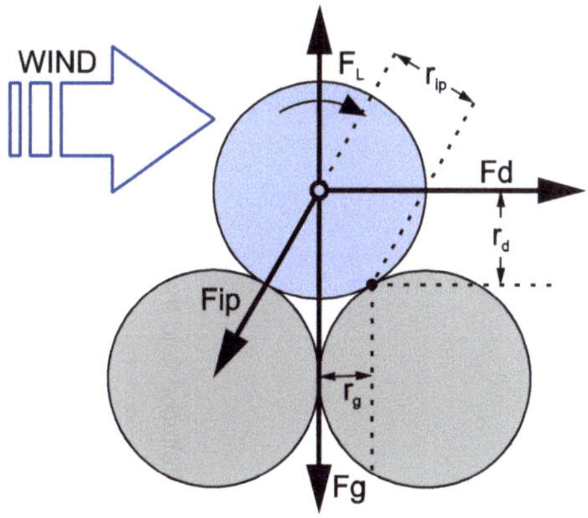

Fig. 1 Motion trigger mechanism described by Shao and Lu (2000)

A particle, initially in contact with others, is displaced by the wind when the moment of the forces exerted on its surface (F_d and F_L) with respect to the support point P equals the moment of the resisting forces (F_g and F_{ip}); that condition is expressed by Eq. (1):

$$r_d F_d \approx r_g \left(F_g - F_L\right) + r_{ip} F_{ip}. \tag{1}$$

Equation (2) is the expression of the threshold friction velocity (u_{*ft}) obtained from Eq. (1) by replacing the forces' mathematical formulas (Bagnold 1941):

$$u_{*ft} = A_{ft} \sqrt{\frac{\rho_p - \rho_a}{\rho_a} g D_p}. \tag{2}$$

where ρ_p and ρ_a are, respectively, the particle and the fluid density, D_p is the particle diameter, g is gravity acceleration, and A_{ft} is a function of the interparticle forces, the suspension forces, and the Reynolds number.

Wind erosion only occurs when the friction velocity exceeds the threshold friction velocity (u_{*ft}). By applying Eq. (2) to a series of experimental data for dissolved sand, Bagnold obtained $A_{ft} = 0.10$ (Bagnold 1941). Using the A_{ft} function proposed by Iversen and White (1982), Shao and Lu (2000) suggested the use of Eq. (3):

$$u_{*ft} = A_N \sqrt{\frac{\rho_p - \rho_a}{\rho_a} g D_p + \frac{\gamma}{\rho_a D_p}}. \tag{3}$$

where γ accounts for the interparticle forces and A_N for the suspension forces and the Reynolds number.

Figure 2 shows the variability of the *threshold friction velocity* as a function of the particle equivalent diameter (Kok et al. 2012). The diagram integrates the research results of various authors: Bagnold (1937), Chepil (1945), Zingg (1953), and Iversen et al. (1976) for sand and dust particles; Fletcher (1976) and Iversen

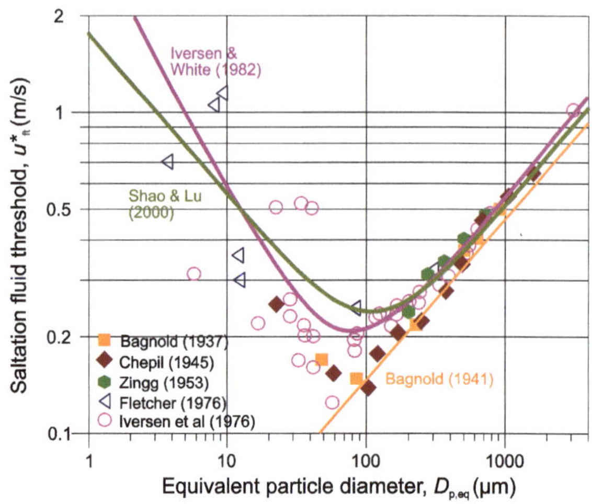

Fig. 2 Saltation fluid velocity as a function of the saltators' size (Kok et al. 2012)

et al. (1976) for other materials. The models proposed by Iversen and White (1982), Shao and Lu (2000), and Bagnold (1941) are also integrated in Fig. 2, while the effect of the particle density is included in the definition of the particle equivalent diameter given by Eq. (4):

$$D_{\mathrm{p,eq}} = \frac{D_{\mathrm{p}}\rho_{\mathrm{p}}}{\rho_{\mathrm{p,sand}}}. \tag{4}$$

where ρ_{p} is the density of the particle and $\rho_{\mathrm{p,sand}}$ is equal to 2650 kgm^{-3}.

Figure 2 shows that larger (D_{p} > 500 μm) and smaller particles (D_{p} < 10 μm) are hardly raised by the wind (i.e., they are raised when the wind takes very high speed), due to their weight in the first case and to the adhesion forces in the second case (Kok et al. 2012). The minimum value of the *threshold friction velocity* (*saltation fluid threshold*) is for particle diameters around 100 μm.

According to Bagnold (1941) and Shao (2008), as the wind speed increases particles with equivalent diameter around 100 μm are lifted in the air; after a short trajectory, they fall onto the surface bouncing several times (*saltation*); the impact with the surface breaks the interparticle bonds releasing smaller particles, which remain suspended in the air, because of their lightness, to be transported by the wind even at considerable distances (*suspension*). The impact of bouncing particles also determines the transfer of momentum to larger particles (between 100 and 500 μm), which are not transported in suspension but move in contact with the surface (*creeping or reptation*).

In line with the above-described conceptual model, wind erosion develops according to the following mechanisms: transport in suspension for long distances (D_{d} < 20 μm), transport in suspension for short distances (20 μm < D_{d} < 70 μm), saltation (70 μm < D_{d} < 100 μm), and creeping or reptation (D_{d} > 500 μm).

Clearly, the attribution of the type of motion to the particle size class depends on the wind speed and is therefore purely indicative.

3 PM Emission

Tailings of metallurgical processes and specifically those deriving from the bauxite treatment (red mud) are composed of very small particles (Type A): 90% of the red mud is typically below 20 μm. Due to the superficial forces, small loose particles (Type A) tend to aggregate to form macro-particles (Type B) with diameter between 20 and 300 μm (Alfaro et al. 1997).

Dust emission from tailing deposits is generated by the following mechanisms:

- direct lifting of loose particles (Type A);
- expulsion of loose particles from the surface due to the impact of macro-particles (Type B), which play the role of saltators (bouncing particles);
- disintegration of bouncing macro-particles (Type B) into loose particles (Type A), as a result of their impact on the surface.

Since the threshold velocity of the macro-particles (Type B) is lower than that of the smaller loose particles (Type A), the emission is triggered by the saltation of macro-particles with equivalent diameter around 100 μm and subsequently, when the wind speed increases, by the lifting of smaller loose particles (Type A) and bigger macro-particles (Type B), according to Fig. 2. The mathematical expressions describing the *saltation threshold velocity* and the dust emission flow refer to the mechanisms described above.

3.1 Threshold Shear Velocity

The general expression of the *saltation threshold velocity* is given by Shao and Lu (2000), with an adjustment that takes into account the effect of the particles' physical characteristics and, in particular, their moisture content (as it determines the onset of interparticle forces that inhibit saltation). Fécan et al. (1999) suggested the use of Eq. (5):

$$
\begin{aligned}
\frac{u_{*\mathrm{wt}}}{u_{*\mathrm{ft}}} &= 1 (w < w'), \\
\frac{u_{*\mathrm{wt}}}{u_{*\mathrm{ft}}} &= \sqrt{1 + 1.2 (w - w')^{0.68}} (w \geq w').
\end{aligned}
\tag{5}
$$

where u_{*wt} is the *threshold friction velocity* for a given moisture content (w); w' is the humidity at which the development of the capillary forces occurs and depends on the clay fraction in the soil (c_s), according to Eq. (6) (Fecan et al. 1999):

$$w' = 0.17c_s + 0.0014c_s^2. \tag{6}$$

3.2 Emitted Dust Flow

The dust flow (kg/m^{2}s) is proportional to the kinetic energy transferred from the saltators to the impact surface. According to Eq. (7), the dust flow is calculated by multiplying the average kinetic energy of the saltators (E_s) by the number of saltators that impact the surface unit in the time unit (n_s), by a given efficiency coefficient (ε) which expresses the mass of dust emitted per unit of kinetic energy transmitted to the impact surface (Eq. 7):

$$F_d = n_s E_s \varepsilon. \tag{7}$$

Many authors (Shao et al. 1993; Duràn et al. 2011; Kok 2010) developed the conceptual relationship expressed by Eq. (7) and suggested the use of Eq. (8) to calculate the dust flow:

$$F_d = C_F \rho_a u_{*it}\left(u_*^2 - u_{*it}^2\right). \tag{8}$$

where C_F is a constant measured in kg/j, u_{*it} is the *impact threshold velocity*, u_* is the *friction velocity*, and ρ_a is the fluid density.

As an alternative to the energetic approach, the emitted dust flow can be estimated as a function of the saltation flow Q, according to Eq. (9):

$$F_d = \alpha Q. \tag{9}$$

On that basis, Marticorena and Bergametti (1995) developed the Eq. (10):

$$F_d = C_K \frac{\rho_a}{g} u_*^3 \left(1 - \frac{u_{*it}^2}{u_*^2}\right)\left(1 + \frac{u_{*it}}{u_*}\right). \tag{10}$$

where the constant C_K has dimensions of m^{-1}.

Shao et al. (1996) developed the *Wind Erosion Assessment Model* (WEAM), according to which the one-dimensional flow of particles with d_d diameter is determined by the saltation of particles with d_s diameter, according to Eq. (11):

$$\hat{F}(d_d, d_s) = \left(\frac{\gamma c_b \rho m_d}{\psi}\right) u_*^3 \left[1 - \left\{\frac{u_{*t}(d_s)}{u_*}\right\}^2\right]. \tag{11}$$

where m_d is the mass of the emitted particles, ψ is the binding energy between the particles and the surface, c_b expresses the efficiency of the saltation bombardment, γ is the dimensionless ratio $(U_0 + U_1)/2 \cdot u_*$, U_0 and U_1 are, respectively, the lifting and the impact velocity of the bouncing particles, and u_* and u_{*t} are, respectively, the friction velocity and threshold friction velocity.

The relationships that express the dependence of the vertical flow from the interparticle forces, the size of individual particles, the presence of crusts, the soil moisture content, and its plastic characteristics are not known in explicit and general terms. The influence of those parameters is instead introduced into the formulas in the form of constant values (*site-specific constants*), which are experimentally determined for each specific case study.

That strategy was followed by the authors of a report regarding the air dispersion modelling of the fugitive dust emitted by the red mud basin managed by Alcoa World Alumina in Australia (Air Assessment for Alcoa World Alumina Australia 2005). The authors of the study have elaborated the PM_{10} flow expression suggested by Shao et al. (1996) by replacing in Eq. (12) the friction velocity u_* and the threshold friction velocity u_{*t}, respectively, with the mean wind velocity u and the threshold velocity ut at 10 m. The parameters that describe the role of the forces acting on the particles (gravity and interparticle forces), the saltators' kinetic energy and the energy transfer efficiency of the bouncing motion (γ, c_b, m_b, ψ) have been incorporated into a single site-specific factor (k). The resulting mathematical model is given by Eq. (12):

$$\begin{aligned} PM_{10} &= k\left[u^3 \cdot \left(1 - \frac{u_t^2}{u^2}\right)\right], \quad \text{per } u > u_t \\ PM_{10} &= 0, \quad \text{per } u \leq u_t. \end{aligned} \tag{12}$$

4 Red Mud Deposits

Red muds are composed of very small and relatively uniform particles (90% under 20 μm) with specific weight in the range between 3.2 and 3.8 g/cm^3. They are characterized by low values of plasticity index and plastic limit (liquid) and are classified as silts. Until the '70s, red muds were disposed in lagoons, with a solid content around 20–25% (wet disposal); currently, mainly to reduce the environmental impact, they are previously dried to a solid content around 55–75% and then disposed in landfills: dry staking deposits, with water content around 55–65%, and dry disposal deposits, with water content around 65–75%. When disposed in lagoons the red mud is a water suspension; in dry staking is a supersaturated solid (water content higher than the liquid limit); in dry disposal is a plastic solid (water content lower than the liquid limit and higher than the plastic limit).

In none of those disposal conditions wind erosion occurs, as it is inhibited by the particle forces that characterize the supersaturated and plastic states. Under dry

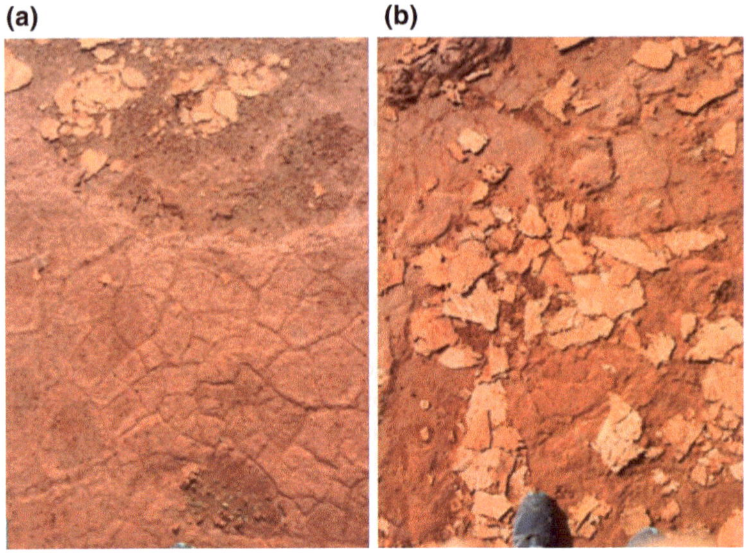

Fig. 3 BRDA surface: **a** mud cracks and particles deposits; **b** particle deposits and crusts

climatic conditions, however, it is possible the undertaking of a drying process that changes the mud into a dried solid: compact rigid crusts a few mm thick are formed at the surface separated by cracks (Fig. 3a), the more widespread and open the higher the initial water content in the mud.

The surface of the crusts does not generate saltators because of the high forces binding the particles inside the dry solid; they are produced instead by:

- the passage of people or vehicles on the surface when the mud is in the plastic state;
- the crushing of dried crusts due to the passage of people or vehicles or mechanical actions (rain, hail, etc.);
- the action of the wind on the crusts' edges (chipping).

Individual particles (type A) and particle aggregates (type B) form a granular material that settles in the mud cracks and in the surface depressions (Fig. 3a, b).

5 Conceptual Emission Model

According to the above description, the following conceptual model can describe the emission of dust from a red mud deposit:

- an initial ON/OFF condition based on the surface water content;
- a subdivision of the basin surface into categories of emitting areas;
- an emission mechanism for each category of emitting areas.

5.1 *ON/OFF Emission Condition and Surface Discretization*

On the basis of the mud water content at the basin surface, it is possible to distinguish:

1. Emission condition (ON), when the mud at the surface is mainly a dry solid ($W < W_p$), so that the surface is formed by the following categories:
   - residual areas in which the mud water content is higher than W_p (Awet);
   - assemblages of A and B particles inside mud cracks and surface depressions (A & B);
   - crust areas (Acrust);
   - crust edges (Achipping);
2. Non-emission condition (OFF), when the mud at the surface is in a plastic state of consistency ($W > W_p$)

5.2 *Emission Mechanism for Each Category of Emitting Areas*

Assemblages A and B particles: the emission from the surface of these deposits follows the model described in Sect. 3.2. The saltation threshold velocity $U_{TA\&B}$ and the vertical flow of dust ($Fd_{A\&B}$) are defined by Eqs. (3) and (12), respectively.

Surface of the crusts: Due to the high particle forces in the dry solid state of the crust, its surface does not produce saltators (except for very high wind speeds). The expulsion of particles from the crust occurs only as a result of the impact of large saltators (over 500 μm) coming from other emitting surfaces and is therefore triggered at a threshold velocity corresponding to that of large saltators (U_{T500}). The vertical flow Fd_{crust} is calculated again with Eq. (12), where k is specific for this type of emission.

Edges of mud crack: The edges of the mud cracks, due to their shape and the fragility of the dry mud, emit particles when they break (chipping). This phenomenon occurs at a threshold wind velocity UT chipping greater than $U_{TA\&B}$. The particle flow is described again by Eq. (12), with an appropriate value of k. As the wind speed increases and exceeds the threshold speed defined for the three categories of areas, the three flows are superimposed.

Once the A and B particles are exhausted, the surface emits only the particles originated from the chipping of the crust edges. If this flow is neglected, the surface can be assimilated to a finite or exhaustible source of particles. In the case of surfaces in which transport and disposal operations take place, the production of new particles is continuous and the surface constitutes an infinite source of dust.

5.3 Preliminary Verification of the Conceptual Model

The described conceptual model has been partially verified for a 1000 × 500 m area within a BRDA. The study was aimed at determining the distribution of the categories of emission areas, the threshold velocities ($U_{TA\&B}$, $U_{Tchipping}$, and U_{Tcrust}), the surface roughness z_0, and the k constants in the relationships that express the vertical flow.

To this end, the following measures were carried out: incidence of the categories of emission areas, wind speed at two different altitudes, particle size distribution of the particle deposits in the BRDA surface, and concentration of PM_{10} upstream and downstream of the emitting surface. Presently, the incidence of the categories of emission areas, the surface roughness z_0, and the value of the $U_{TA\&B}$ threshold speed have been determined.

An area of 200 m in the direction normal to the wind and 500 m in the direction parallel to the wind has been identified on the BRDA surface. The incidence of the categories of emission areas was determined by dividing this area into 10 m wide and 500 m long longitudinal strips and each strip in 10 m long segments; for each resulting 10 × 10 m^2, the incidence of the three categories of areas was evaluated. The resulting average values were wet area 50%; crust area 40%; area with A and B particle deposit 9.8%; and chipping area 0.2%.

The roughness z0 was obtained from the contemporary measurements of the wind velocity at two heights (u_1 at z_1 = 2.0 m and u_2 at z_2 = 7.0 m from the basin surface), in conditions of neutral atmospheric stability. Figure 4 shows the trend of the instantaneous wind speed (1 measure every 5 s) and the mean values over ½ h. A mean value of 0.42 mm was calculated with Eq. (13):

$$z_0 = e^{\left(\frac{u_1 \ln(z_2) - u_2 \ln(z_1)}{u_1 - u_2}\right)}. \tag{13}$$

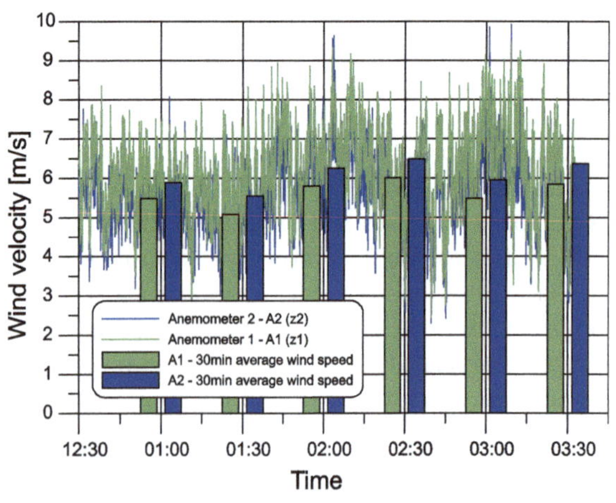

Fig. 4 Instantaneous wind speed (1 measure every 5 s) and mean values over ½ h

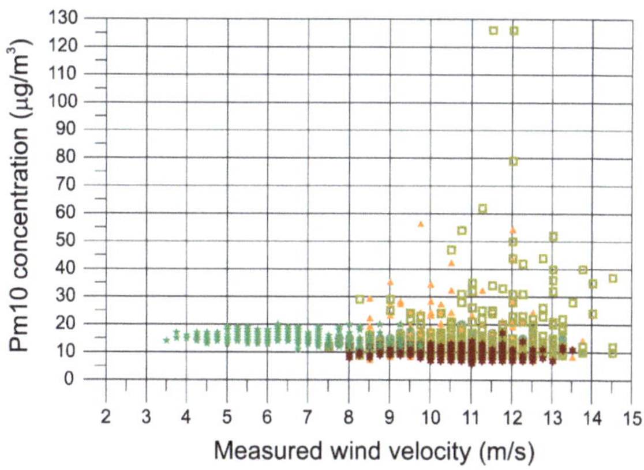

Fig. 5 PM10 concentrations for each wind velocity value (referred at 10 m of height)

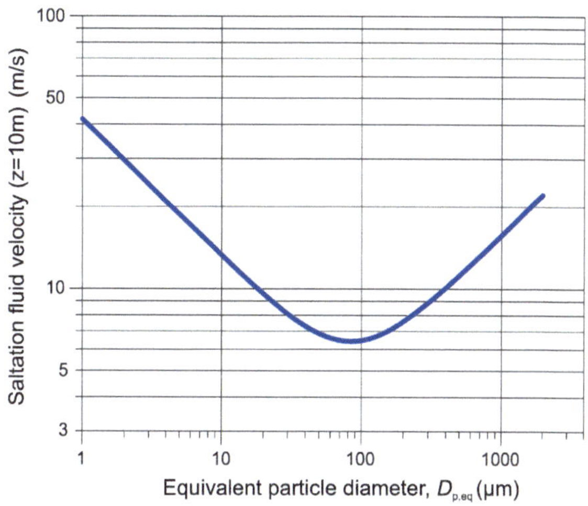

Fig. 6 Saltation velocity at 10 m of height as a function of the saltators' size

The threshold velocity $U_{TA\&B}$ was determined from the PM_{10} concentration measured at the four stations located downwind of the 200 × 500 m emission area. The obtained $U_{TA\&B}$ was used to have an indication of the saltators' size.

Figure 5 shows the measured PM_{10} concentrations for each wind velocity (measured at 10 m) and indicates that PM_{10} concentration exceeds the background value when the wind speed at 10 m exceeds 8 m/s. This value represents in fact the threshold velocity $U_{TA\&B}$ (the lowest threshold velocity among the three defined emission modes). The size of the saltators that triggers the observed emission is deduced from the Shao formula (3), expressed as shown in Fig. 6, for the threshold velocity of 8 m/s: It results in the range between 30 and 200 µm.

This result is consistent with the particles size distribution of the material in the A&B type areas, where particles under 250 µm are about 27.5% (Table 1).

Table 1 Size distribution of the types A and B particles deposits

Particle size (mm)	+2.0	−2.0 +1.0	−1.0 +0.5	−0.5 +0.25	−0.25
Weight (%)	28.916	13.222	13.864	16.488	27.508

6 Conclusions

This article deals with the emission of PM from the surfaces of the bauxite residue disposal areas (BRDA) exposed to wind erosion. In fact, the examination of the technical and scientific reports has shown that the emission factors proposed for other types of erodible surfaces cannot be directly applied to those deposits, because of their peculiar characteristics: wide and flat surfaces with low roughness and residue physical state dependent on its moisture content.

The action of the wind over the BRDA surfaces has been studied with the aim of developing a conceptual model capable of predicting the conditions that trigger the emission of dust and the emitted flux. The model is based on the observation that the emission occurs only if the mud at the basin surface presents a dry solid physical state and includes three different emission mechanisms related to the presence of loose particle deposits, stiff crusts and mud cracks. The overall potentially emissive surface is first divided into categories of emissive areas; each category is characterized by a set of values, which accounts for the areal extent, the threshold velocity and a k value (parameters included in the flux formula). The overall emission flux can be calculated as the sum of the emission fluxes of each single category of areas for which the threshold velocity and the emission flux have been determined.

The model was applied to a specific case study located in Sardinia. A first experimental phase included the division of the basin surface into categories of emission areas, the calculation of the threshold velocity for the particle deposits ($U_{TA\&B}$) and the calculation of the saltators size. The successive experimental phase is currently under development and includes on site measurements of PM_{10} concentration with higher wind speeds, aimed at evaluating the threshold velocity and the k factors for the other emission mechanisms (crust emission and edge chipping).

Acknowledgements Investigation carried out in the framework of projects conducted by CINIGeo (National Inter-university Consortium for Georesources Engineering, Rome, Italy). "RE-MINE -REstoration and remediation of abandoned MINE sites", funded by the Fondazione di Sardegna and Regional Sardinian Government (Grant CUP F72F16003160002).

References

Air Assessment for Alcoa World Alumina Australia: Air dispersion modelling of fugitive emissions. Wagerup Refinery (2005)

Alfaro, S.C., Gaudichet, A., Gomes, L., Maille, M.: Modeling the size distribution of a soil aerosol produced by sandblasting. J. Geophys. Res. **102**, 11239–11249 (1997)

Bagnold, R.A.: The transport of sand by wind. Geograph. J. **89**, 409–438 (1937)

Bagnold, R.A.: The Physics of Blown Sand and Desert Dunes. Methuen, New York (1941)

Chepil, W.S.: Dynamics of wind erosion: II. Initiation of soil movement. Soil Sci. **60**, 397–411 (1945)

Duràn, O., Claudin, P., Andreotti, B.: On aeolian transport: grain-scale interactions, dynamical mechanisms and scaling laws. Aeolian Res. **3**, 243–270 (2011)

Fecan, F., Marticorena, B., Bergametti, G.: Parametrization of the increase of the aeolian erosion threshold wind friction velocity due to soil moisture for arid and semi-arid areas. Ann. Geophys. **17**, 149–157 (1999)

Fletcher, B.: Incipient motion of granular materials. J. Phys. D Appl. Phys. **9**, 2471–2478 (1976)

Iversen, J.D., White, B.R.: Saltation threshold on Earth. Mars Venus Sedimentol **29**, 111–119 (1982)

Iversen, J.D., Pollack, J.B., Greeley, R., White, B.R.: Saltation threshold on Mars—effect of interparticle force, surface-roughness, and low atmospheric density. Icarus **29**, 381–393 (1976)

Kok, J.F.: An improved parameterization of wind-blown sand flux on Mars that includes the effect of hysteresis. Geophys. Res. Lett. **37**, L12202 (2010)

Kok, J.F., Parteli, E.J.R., Michaels, T.I., Bou, D.: Karam: the physics of windblown sand and dust. Rep. Prog. Phys. **75**, 106901 (2012)

Marticorena, B., Bergametti, G.: Modeling the atmospheric dust cycle: 1 Design of a soil-derived dust emission scheme. Geophys. Res. **100**, 16415–16430 (1995)

Shao, Y.P.: Physics and Modelling of Wind Erosion, 2nd edn. Springer), Heidelberg (2008)

Shao, Y.P., Lu, H.: A simple expression for wind erosion threshold friction velocity. J. Geophys. Res. **105**, 22437–22443 (2000)

Shao, Y., Raupach, M.R., Findlater, P.A.: Effect of saltation bombardment on the entrainment of dust by wind. J. Geophys. Res. **98**, 12719–12726 (1993)

Shao, Y.P., Raupach, M.R., Leys, J.F.: A model for predicting aeolian sand drift and dust entrainment on scales from paddock to region. Aust. J. Soil Res. **34**, 309–342 (1996)

Zingg, A.W.: Wind tunnel studies of the movement of sedimentary material. In: Proceedings of 5th Hydraulic Conference, University of Iowa, Iowa City, pp. 111–135 (1953) (Studies in Engineering Bulletin 34)

Ash, Volatile Matter and Carbon Content Influence on Spontaneous Combustion Liability of Coal-Shales

M. Onifade and B. Genc

1 Introduction

The event of spontaneous combustion is one of the major problems in the coal value chain (Liu and Zhou 2010; Onifade and Genc 2018a, b; Phillips et al. 2011). It is well-known that coals and coal-shales undergo self-heating when exposed to oxygen in the air (Onifade and Genc 2018b, c; Onifade et al. 2018). The heat produced may become faster than the heat dissipated to the surrounding (Kim and Chaiken 1990; Dullien 1979). The first mechanism of spontaneous combustion involves heat produced during the physical and chemical processes, which play a significant role in the increase of coal temperature until the ignition point, and the second mechanism involves the reaction of oxygen with these materials either as a system of successive oxidation reactions or in the form of a single constant process (Onifade et al. 2018). Most of the heat transferred may be caused by conduction to surrounding strata (Akande and Onifade 2013; Akande et al. 2013). Spontaneous combustion of coal-shales and spoil heaps fire may result from the accumulation of heat generated and heat dissipated to the surroundings (Onifade and Genc 2018a; Phillips et al. 2011). Spontaneous combustion in waste dumps, spoil heaps and coal-shales is comparable to coal oxidation in which carbon oxide and carbon dioxide are produced (Onifade and Genc 2018b; Onifade et al. 2018; Kim and Chaiken 1990; Restuccia et al. 2017; Rumball et al. 1986).

Coals and coal-shales which contain variable proportions of organic matter (macerals) and inorganic materials (mainly crystalline) may undergo self-heating (Onifade and Genc 2018c; Restuccia et al. 2017; Rumball et al. 1986). This enables the rock to be porous to air, and with the increased surface area, the organic particles have reactive oxidation sites (Dullien 1979). The reasons for coal-shales to

M. Onifade · B. Genc (✉)
The School of Mining Engineering, University of the Witwatersrand,
Johannesburg, South Africa
e-mail: bekir.genc@wits.ac.za

E. Widzyk-Capehart et al. (eds.), *Proceedings of the 18th Symposium on Environmental Issues and Waste Management in Energy and Mineral Production*,
https://doi.org/10.1007/978-3-319-99903-6_3

Fig. 1 **a** Burning spoil heaps at Tweefointein Mine and **b** Symptoms of shale self-heating at Goedgevonden Colliery, Witbank, South Africa (Onifade and Genc 2018b, c; Onifade et al. 2018)

undergo spontaneous combustion could be caused by the volume of pyrite, organic composition, nature and rank of the associated coal (Restuccia et al. 2017; Rumball et al. 1986; Mastalerz et al. 2010). Spontaneous combustion of coal-shales has been witnessed in Witbank coalfield, South Africa to cause self-heating in selected bands of coal seams (above and below), spoil heaps, waste dumps and highwalls (Onifade and Genc 2018b, c; Onifade et al. 2018) (Figs. 1 and 2). Coal seam and spoil heaps comprise of weathered coal, clays, pyritic shales, coal-shales, slate and other strata associated with the coal seam (Onifade and Genc 2018b; Onifade et al. 2018; Kim and Chaiken 1990; Restuccia et al. 2017; Rumball et al. 1986). Limited research has been carried out to understand the characteristics of coal-shales exposed to an oxidizing environment (Onifade and Genc 2018b, c,d; Onifade et al. 2018; Restuccia et al. 2017; Rumball et al. 1986). Many studies have been carried out on

Fig. 2 Symptoms of self-heating at mine face at Khwezela Mine (Bokgoni Pit), Witbank, South Africa (Onifade and Genc 2018b, c, d; Onifade et al. 2018)

spontaneous combustion liability of coals both experimentally and computationally (Genc et al. 2018; Genc and Cook 2015; Gouws and Wade 1989a, b; Onifade and Genc 2018; Stracher and Taylor 2004). However, no sufficient information has been reported on spontaneous combustion liability of coal-shales.

This paper presents the results of some intrinsic properties (ash, volatile matter and carbon content), and spontaneous combustion tests carried out on selected coal-shales in South African coal mines. This will be suitable to identify the influence of these intrinsic factors on spontaneous combustion liability and provides better understanding in predicting the initiation of coal-shale fires in coal mines.

2 Characterization and Spontaneous Combustion Tests

2.1 Sample Collection

The coal-shales used in this research were gathered (using ply sampling technique) from three different coal mines, South Africa, and to avoid further oxidation, collected samples were kept in airtight bags. Each ply sample bag was individually labelled. Six samples representative of in situ coal-shales were collected from areas known to be self-heating (highwalls and selected bands between coal seams). Characterization tests (proximate and ultimate analysis) and spontaneous combustion liability test (Wits-CT tests) were conducted on the coal-shales.

2.2 Sample Characterization

The sample lumps were reduced with the use of a crusher and ball mill to −250 μm for geochemical and −212 μm for spontaneous combustion tests. Similar preparation is related to studies reported by Onifade and Genc (2018b) and (c). The determination of the volatile matter, ash and carbon content of samples was carried out according to the American Society for Testing and Material Standards (ASTM) (ASTM, D-3173-17a 2017; ASTM, D-3174-11 2011; ASTM, D-3175-17 2017). Fixed carbon was obtained by subtracting the sum of the percentage of volatile matter, ash and moisture from 100. The carbon content was determined using a LECO TruSpec CHNS analyzer (Fig. 3) after calibration with sulfamethazine based on the International standard organization (ISO) (2001). The results pertaining to the coal-shales are discussed therein, and detail description on the characteristics of the samples is extensively described in the study by Onifade and Genc (2018b).

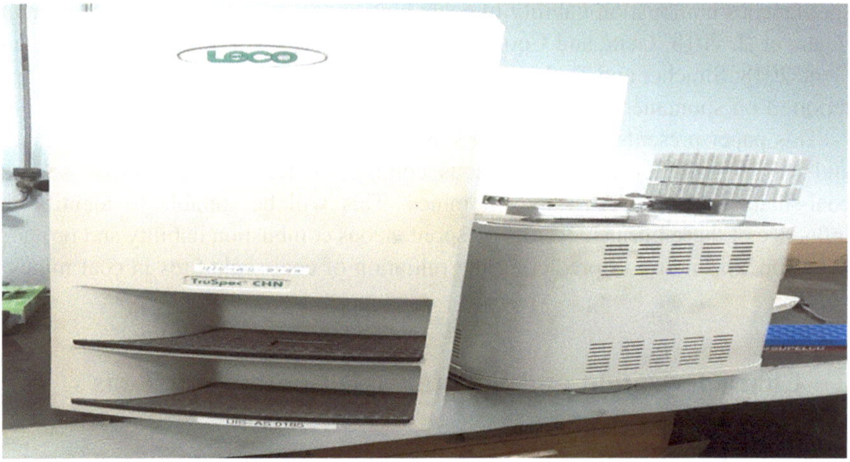

Fig. 3 LECO TruSpec CHN analyzer

2.3 *Wits-CT Tests*

The Wits-Ehac index was developed at the University of the Witwatersrand to measure the spontaneous combustion liability of coal. However, when testing some coal-shales, in some cases the Wits-Ehac index failed to produce results due to the low reactivity of the tested samples. To overcome this problem, a new equipment was developed to test coal and coal-shale undergoing chemical reactions with air, and a new index is obtained. This new index called the Wits-CT index (Onifade et al. 2018). The equipment has been used to test the spontaneous combustion liability of coals and coal-shales (Onifade and Genc 2018a, b, c, d; Onifade et al. 2018). The details of the experimental procedure are reported in study by Onifade et al. (2018). An illustration of the experimental setup is indicated in Fig. 4. The index is calculated from the formula in Eq. 1 (Onifade et al. 2018).

$$\text{Wits} - \text{CT Index} = (\text{TM}/24 + \text{TR}) * \%\,\text{Cad} \qquad (1)$$

where: TM is the difference between the sum of maximum temperatures of each thermocouple in the autoclave and room temperature (22 °C), TR is the difference between the peak temperature and initial temperature during oxidation reaction in degree Celsius, %Cad is the air-dried percentage of carbon content of the sample, * is a multiplication sign, and 24 is the test duration and is constant.

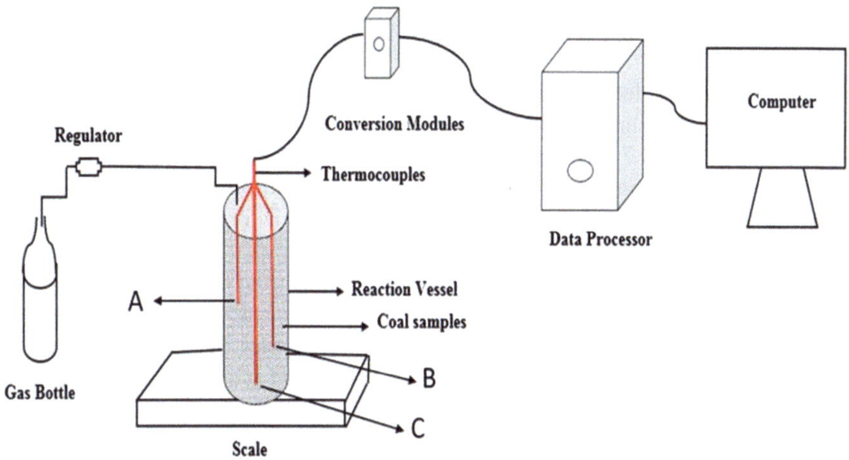

Fig. 4 Schematic of the Wits-CT apparatus (Onifade et al. 2018)

3 Results and Discussion

The results for the geochemical data (proximate and elemental analysis) and spontaneous combustion tests obtained from six coal-shale samples are presented in Table 1. The coal-shales are rich in mineral matter as indicated by the high ash content (Table 1). All samples are hence classified as carbonaceous shales (ash above 50 wt%). Sample SN has the lowest ash content, highest carbon content and highest spontaneous combustion liability index.

Table 1 Results of ash, volatile matter, carbon content (wt%, ad) and spontaneous combustion tests for coal-shale samples

Samples	Moisture	Ash	Volatile matter	Fixed carbon	Carbon	Wits-CT index
SH	0.8	88.7	8.5	2	2.66	0.27
SG	0.8	84.3	10.7	4.2	6.02	0.67
SA	1.4	78.5	11.2	8.9	11.5	1.33
SK	1	79.1	11.7	8.2	9.75	1.18
SM	0.8	76.9	11.7	10.6	12.5	1.44
SN	1.5	51.5	16.6	30.4	33.7	3.99

3.1 Results

The results of the selected intrinsic factors and spontaneous combustion tests carried out on the coal-shale samples are presented in Table 1. The ash content ranges between 51.5 and 88.7 wt%, respectively. The cause of the high ash content is due to the peat depositional environment where the condition of flooding of the paleomire occurred periodically during deposition. This is in line with the studies reported by Onifade and Genc (2018b), Restuccia et al. (2017), Rumball et al. (1986) and Mastalerz et al. (2010). It was observed that the ash contents vary from one another between bands of coal seams. Sample SN has the lowest ash content among the samples analyzed. The ash content of samples SH and SG is considerably high compared to the other samples. Sample SH with the highest ash contents has the lowest liability index, while sample SN samples with the lowest ash content has the highest spontaneous combustion liability index. The low and high liability index of the samples could be due to the influence of minerals absorbing heat within the coal-shales. Similar studies are reported by Onifade and Genc (2018b, d). It was found that the spontaneous combustion liability of coal-shale decrease with increasing ash content and vice versa. This characteristic is similar to those exhibited by coals in studies reported by Onifade and Genc (2018b, d). Therefore, coal-shale samples display similar ash behaviour as coal with respect to spontaneous combustion. The air-dried volatile content for the samples varies between 8.5 and 16.6 wt%, respectively. Sample with the highest volatile matter content has high spontaneous combustion liability index, while sample with low volatile matter has low liability index. Similar studies are reported by Onifade and Genc (2018b) and Restuccia et al. (2017). Coal-shale SN has the highest volatile matter content than the other samples. It was found that coal-shale with high volatile matter content is more liable to spontaneous combustion. This characteristic of coal-shale with high volatile matter content corresponding to high liability index is similar to coals with high volatile matter content (Onifade and Genc 2018b, d).

The determined air-dried carbon content varies between 2.66 and 33.7 wt%, respectively. Sample SH has the lowest carbon and the lowest liability index, while sample SN with the high carbon content has the highest spontaneous liability index. It was observed that the analyzed coal-shale samples have low carbon content when compared with some coals. Similar studies are reported by Onifade and Genc (2018b, d), Restuccia et al. (2017) and Rumball et al. (1986). It is more likely to suggest that the ability of coals, coal-shales and other carbonaceous materials to undergo self-heating is directly related to the amount of organic and inorganic matter present within them. If the self-heating rate of carbonaceous materials is related with the heat generated and heat dissipated from a single process, it is expected that the rise in temperature increases the reaction rate of that process. The carbon contents found in the coal-shales varied from one sample to the other. The results indicate that different reactions with different activation energies take place during spontaneous combustion. Coal-shale SH has the lowest Wits-CT, and coal-shale SN has the highest Wits-CT index among the coal-shale samples.

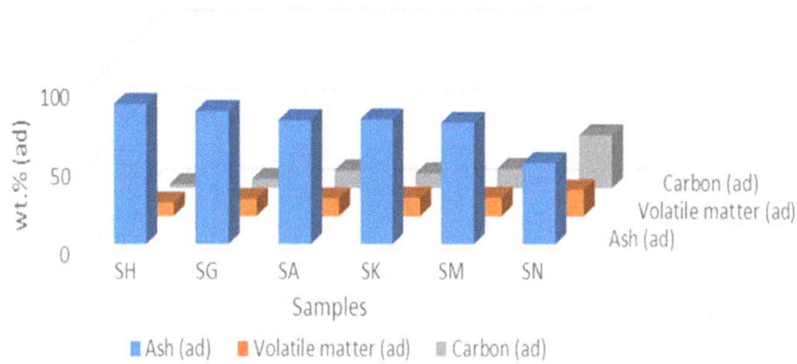

Fig. 5 Representation of ash, volatile matter and carbon content of coal-shale samples

The study shows that high-risk coal-shales have high Wits-CT value. The results show that the event of spontaneous combustion in coal mines is caused by a number of organic and inorganic constituents within coal-shale. Similar study is reported by Mastalerz et al. (2010). Coal-shales undergo self-heating when they absorb sufficient oxygen and moisture under the influence of atmospheric conditions. The heat generated due to the influence of oxygen exceeds the heat dissipated to the surrounding (via conduction, convection and radiation) and accumulates to cause spontaneous combustion (Onifade and Genc 2018b, d). Figure 5 shows that coal-shales with lower ash content, high carbon and high volatile matter are more liable to spontaneous combustion. Therefore, the influence of these factors may play a significant role in evaluating the occurrence of spontaneous combustion.

The decreasing ash content affects the self-heating potential by causing high spontaneous combustion liability index for the coal-shales as shown in Fig. 6a. Similar studies are reported by Onifade and Genc (2018c, d). Figure 6a shows that coal-shales with high ash content are showing low liability index, while coal-shales with low ash content seem to have high spontaneous combustion liability index. When the effect of the ash content is considered on linear regression analysis, it is negative on spontaneous combustion liability index. Similar studies are reported by Onifade and Genc (2018c, d). The spontaneous combustion liability for the coal-shales was found to increase with increasing volatile matter as shown in Fig. 6b. The results obtained from the linear regression analysis showed a positive effect on the self-heating potential. Therefore, as the volatile matter increases, the oxidation capacity of coal-shale increases in general. Similar studies are reported by Onifade and Genc (2018c, d). The sharp fall in the linear graph (Fig. 6b) could be caused by high mineral matter, which might act as heat absorbing capacity of the mineral within the coal-shale and hence suppress the self-heating rate. The increasing carbon content affects the self-heating potential of the coal-shales by causing a high spontaneous combustion liability as shown in Fig. 6c. The study indicated that as the carbon content increases, the spontaneous combustion liability

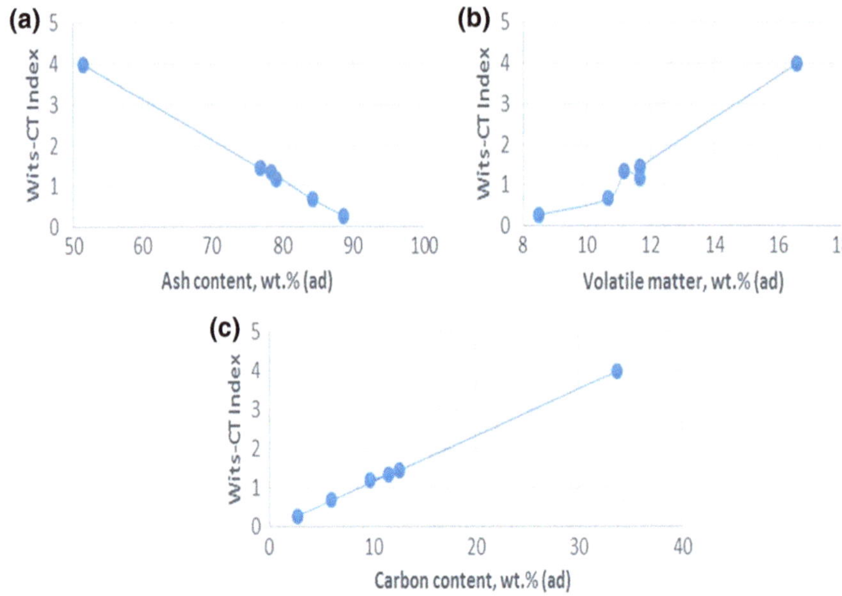

Fig. 6 Effects of **a** ash content **b** volatile matter content **c** carbon content on Wits-CT index

index of coal-shales increased. The results obtained from the linear regression analysis indicated a positive linear effect on the self-heating potential. Similar studies are reported by Onifade and Genc (2018c, d).

4 Conclusion

In this paper, the influence of carbon, volatile matter and ash contents on spontaneous combustion liability of coal-shales was investigated. The Witbank coal-shales were assessed using the Wits-CT spontaneous combustion liability index. It was found that ash shows a negative effect on the spontaneous combustion liability, while volatile matter and carbon showed positive effects with the self-heating potential. Coal-shales with high volatile matter and carbon content and low ash content are more liable to spontaneous combustion liability index and vice versa. The study indicated that the intrinsic properties evaluated affect spontaneous combustion liability of coal-shales and may be used as a reference in predicting their likelihood towards coal-shale spontaneous combustion.

Acknowledgements The authors would like to thank the staff of Glencore and Anglo American coal mines for their help during sample collection. The authors wish to express gratitude to Coaltech and Julian Baring Scholarship Fund (JBSF) for their financial support. The work presented here is part of a Ph.D. research study in the School of Mining Engineering at the University of the Witwatersrand.

References

Akande, J.M., Onifade, M.: Modeling of Okaba underground coal mining ventilation system. Int. J. Eng. Technol. **3**(7), 766–772 (2013)

Akande, J.M., Onifade, M., Aladejare, A.E.: Determination of airflow distributions in Okaba underground coal mine. J. Min. World Expr **2**(2), 40–45 (2013)

ASTM, D-3174-11. Standard test method for ash in the analysis sample of coal and coke from coal and standard classification of coals by rank. ASTM International, West Conshohocken, PA (2011). www.astm.org

ASTM, D-3175-17. Standard test method for volatile matter in the analysis sample of coal and coke. ASTM International, West Conshohocken, PA (2017). www.astm.org

ASTM, D-3173-17a. Standard test methods for moisture in the analysis sample of coal and coke. ASTM International, West Conshohocken, PA (2017). www.astm.org

Dullien, F.: Porous Media Fluid Transport and Pore Structure, p. 79. Academic Press (1979)

Genc, B., Cook, A.: Spontaneous combustion risk in South African coalfields. J. South. Afr. Inst. Min. Metall. **115**, 563–568 (2015)

Genc, B, Onifade. M, and Cook.: A Spontaneous combustion risk on South African coalfields: part 2. In: Proceedings of the 21st International Coal Congress of Turkey 'ICCET' April 11–13, 2018, Zonguldak, Turkey, pp. 13–25 (2018)

Gouws, M.J., Wade, L.: The self-heating liability of coal: predictions based on simple indices. Mining Science and Technology **9**, 75–80 (1989a)

Gouws, M.J., Wade, L.: The self-heating liability of coal: predictions based on composite indices. Min. Sci. Technol. 9, 81–85 (1989b)

ISO 12902:2001: Determination of total carbon, hydrogen and nitrogen

Kim, A.G., Chaiken, R.F.: Relative self-heating tendencies of coal, carbonaceous shales and coal refuse. In: West Virginia, Paper presented at the 1990 Mining and Reclamation conference and Exhibition, Charleston, West Virginia, pp. 535–542 (1990)

Liu, L., and Zhou. F.B.: A comprehensive hazard evaluation system for spontaneous combustion of coal in underground mining. Int. J. Coal Geol. **82**, 27–36 (2010)

Mastalerz, M., Drobniak, A., Hower, J. C., O'keefe, J.M.K.: Spontaneous combustion and coal petrology. In: Stracher, G.B., Sokol, E.E., Prakash, A (eds) Coal and Fires: A Global Perspective. Coal-Geology and Combustion **1**, 47–62 (2010)

Onifade, M., Genc, B.: A review of spontaneous combustion studies—South African context. Int. J. Min. Reclam. Environ. 2–17 (2018a)

Onifade, M., Genc, B.: Spontaneous combustion of coals and coal-shales. Int. J. Min. Sci. Technol. (2018b) https://doi.org/10.1016/j.ijmst. 2018.05.013

Onifade, M., Genc, B.: Modelling spontaneous combustion liability of carbonaceous materials. Int. J. Coal Sci. Technol. (2018c). https://doi.org/10.1007/s40789-018-0209-2

Onifade, M., Genc, B.: Prediction of the spontaneous combustion liability of coal and coal-shale using statistical analysis. In: Society of Mining Professors, 6th Regional Conference, March 12–13, Johannesburg, South Africa, 2018, pp. 63–82 (2018d)

Onifade, M., Genc, B.: Establishing relationship between spontaneous combustion liability indices. In: Proceedings of the 21st International Coal Congress of Turkey 'ICCET' 11–13 April 2018, Zonguldak, Turkey, 1–11 (2018e)

Onifade, M., Genc, B., Carpede, A.: A new apparatus to establish the spontaneous combustion propensity of coals and coals. Int. J. Min. Sci. Technol. https://doi.org/10.1016/j.ijmst. 2018. 05.012 (2018)

Phillips, H., Chabedi, K., Uludag, S.: Best practice guidelines for South African Collieries, 1–129 (2011)

Restuccia, F., Ptak. N., Rein. G.: Self-heating behaviour and ignition of shale rock. Combustion and Flame **176** 213–219 (2017)

Rumball, J.A., Thomber, M.R., Davidson. L.R.: Study of chemical reactions leading to spontaneous combustion of pyritic black shale at MT Whaleback, Western Australia. In: Western Australia, Symposia series, Australasian Institute of Mining and Metallurgy, pp. 133–139 (1986)

Stracher, G.B., Taylor, T.P.: Coal fires burning out of control around the world: thermodynamic recipe for environmental catastrophe. Int. J. Coal Geol. **59**(1), 7–17 (2004)

Effect of the Heat Input by Dolerite Intrusions and the Propensity for Spontaneous Combustion in the Highveld Coalfields, South Africa

E. R. Mokone, T. Zvarivadza and F. Sengani

1 Introduction

Spontaneous combustion of coal has been a major problem for the industry (both coal producers and consumers), human health and the environment worldwide. Coal is a carbonaceous rock composed of combustible material of carbonized organic matter that may be oxidized at different conditions, especially when it comes into contact with the atmosphere and moisture (Avila et al. 2014; Pone et al. 2007; Sasaki and Sugai 2011). Put in simple terms, the combustible nature of coal arises from its formation processes in which, during the carboniferous period, the plant's biomass underwent burial metamorphism in anaerobic and aquatic environments (Speight 2005). Many authors believe that spontaneous combustion of coal is further elevated by the emplacement of igneous intrusions resulting in chemical and thermal alteration of the coal producing mainly methane, carbon dioxide and organic compounds (Golab 2003). The combined effects of the spontaneous combustion of coal and igneous intrusions are detrimental to coal mining affecting the quality of coal, atmospheric pollution by gaseous by-products, safety, productivity and economic feasibility (Avila et al. 2014; Golab 2003).

The burning of coal is a great concern globally as the by-products may cause major environmental problems. Analyses of gas from coal fires in India, USA, China, South Africa and other countries show that there are toxins harmful to human health in coal-fire gas that cause carcinogenic proprieties such as benzene,

E. R. Mokone
Johannesburg, South Africa

T. Zvarivadza · F. Sengani (✉)
School of Mining Engineering, The University of the Witswatersrand, Johanneburg, South Africa
e-mail: fhatugeorge@gmail.com

T. Zvarivadza
e-mail: tawanda.zvarivadza@wits.ac.za

E. Widzyk-Capehart et al. (eds.), *Proceedings of the 18th Symposium on Environmental Issues and Waste Management in Energy and Mineral Production*,
https://doi.org/10.1007/978-3-319-99903-6_4

xylene and toluene (Finkelman 2004; Pone et al. 2007; Snyman and Barclay 1989). Moreover, the Intergovernmental Panel on Climate Change identified the uncontrollable spontaneous combustion of coal as a potential source emitting greenhouse gases (Pone et al. 2007). Having listed some of the detrimental effects caused by spontaneous combustion of coal, it is necessary to conduct more research to establish criteria necessary for future environmental impact assessment and regulations.

Usually, analyses and characterization of the metamorphic effects on coal are represented in terms of proximate and ultimate analyses, e.g. the in-house study by (Denisov and Metelitsa 1968; Emsbo-Mattingly and Stout 2011) and many others. The proximate analysis of coal is a measure of products determined from heating of the coal conveyed in four parameters: fixed carbon, volatile matter, moisture and ash content (Speight 2005). In contrast to the proximate analysis, there is ultimate analysis, which provides information on the elemental composition of coal. Thus, ultimate analysis includes the weight percentage determination of carbon, hydrogen, oxygen, nitrogen, sulphur and other trace elements that occur within coal (Speight 2005). However, the approach implemented in this study utilizes solely the volume of aromatic and aliphatic organic compounds on the coal surface to investigate their effect on the propensity of spontaneous combustion of coal.

This study investigates and documents the impact of the heat input (metamorphic effect) by the emplaced dolerite intrusions of the Karoo age in the 2 Secunda area Highveld coalfields, South Africa. The main aim of this research is to evaluate the chemistry of the organic compounds in the coal samples, thus tracing changes in chemistry within the coal material and consequently determining the propensity for spontaneous combustion triggered by the intrusive. To achieve this aim, analyses were performed on two samples of coal representative of the positions proximal and distal to the dolerite intrusion.

2 Research Approach

Principal objective of this study was to utilize chemistry data of the coal samples from a coal seam in the Highveld-Secunda coalfield in the vicinity of dolerite intrusions. The coal samples were collected from a coal seam located in the Secunda area. Sampling focused entirely on the C4 lower coal seam at the Ithembalethu shaft of Sasol Mining Secunda. The two coal samples collected represent the two proximal and distal positions in relation to their distance from the 3-m-wide dolerite sill. One coal sample represented a direct contact with the dolerite intrusion (0 m) and the other sample represented coal further away (20 m) from the dolerite intrusion contact.

The samples of this study were investigated using a combination of techniques. The collected coal samples were pulverized and homogenized through grinding to a powder of <1 mm size dimension. The coal powder then went through accelerated solvent extraction (ASE) with toluene as the extracting solvent in order to detect and extract all the organic compounds within the homogenized coal. The ASE allows a liquid concentrate composed of organic compounds to be created from solid particles. The gas chromatographic analysis of the organic-rich extraction of the coal sample was then performed on Pegasus® 4D GC x GC-TOFMS in order to characterize and quantify the organic compounds in the coal sample. The data output of the results is in terms of excel spreadsheets that are utilized in this study for statistical analyses purposes.

The chemical data obtained from the chromatographic analyses was further analysed and arranged in a simplistic manner. The data was presented in terms of bar graphs that revealed and compared concentrations (in area percentage) of different components. It should be pointed out that the 'area-%' of the organic compounds refers to the chemical identification of different peak series obtained from the chromatographic analyses. However, for the purpose of this study, although the different peaks do not represent the true concentrations of the different compounds on the coal surface, the measure of the peaks from the analyses should provide an estimate of the amount of the present compounds in the coal material. The data output of the proximal coal to the intrusion reported 1406 different organic compound peaks, whereas the distal coal to the intrusion reported 341 peaks.

The proximal and distal chemical data were treated separately and further subdivided into groups based on their functional groups and dominance. These components were further attributed into two major groups of importance, namely aliphatic and aromatic compounds, by assessing the chemical structures of each and every individual component based on the organic chemistry knowledge. The aliphatic and aromatic compounds resulted in a total of 16 and 41 subgroups, respectively. However, there was a presence of 27 unknown data points of organic compounds determined from the proximal coal. These unknown compounds were not allocated to any of the major groups.

3 Results of the Study

The organic compounds presented in the analysed coal samples consisted of a variety of functional groups that were classified into principal groups of aliphatic and aromatic compounds. The results of all analysed organic compounds in proximal and distal coal samples are presented in Figs. 1, 2, 3 and 4.

The analyses of the proximal coal sample (Fig. 1) show that the coal that is assumed to have received the maximum heat input is composed of more aromatic (~76 area-%) than aliphatic (~23 area-%) compounds and also a low concentration of unclassified unknown compounds (~0.2 area-%). In contrast, the distal coal sample analyses (Fig. 2) indicate that the composition of the organic compound is dominated by aliphatic compounds at ~65 area-% and the remaining concentration of ~35 area-% is made up of aromatic compounds. The distal coal sample contributes 73.4%, and the proximal coal sample contributes 26.6% of the overall aliphatic compounds. In addition, the contributions of proximal and distal coal samples to the overall aromatic compounds composition are 68.3% by the former and 31.7% by the latter.

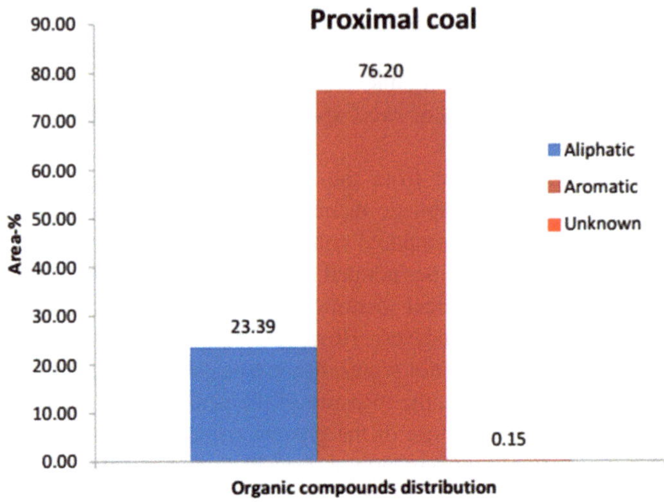

Fig. 1 Concentrations (in area-%) of organic compounds in proximal coal (direct contact) to the intrusion

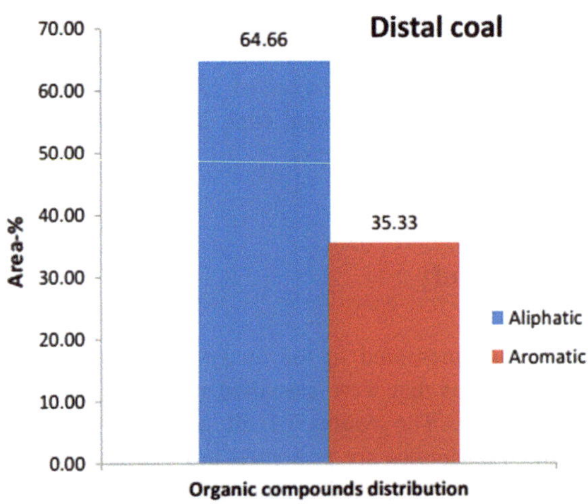

Fig. 2 Concentrations (in area-%) of organic compounds in distal coal (20 m away) to the intrusion

The comparisons among the 16 subgroups of the aliphatic compounds are shown in Fig. 3. *n*-Alkanes (straight-chain alkanes) are part of the organic compounds that make up most of the overall aliphatic compositions in both proximal and distal coal samples. These *n*-alkanes include octane, decane and higher alkanes (*n*C20–*n*-C31), all contributing ~12.2 area-% in proximal coal and ~26.1 area-% in distal coal.

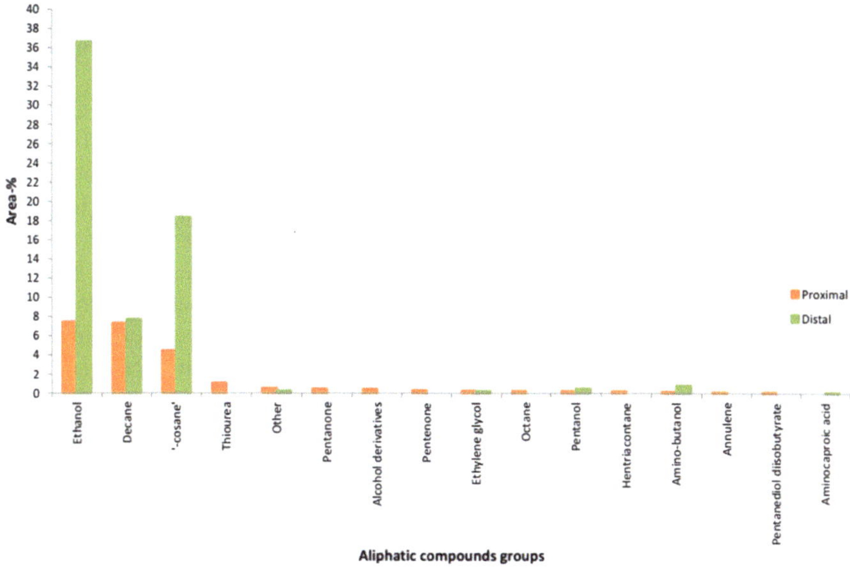

Fig. 3 Comparison of individual groups of aliphatic organic compounds concentrations (in area-%) in relation to the distance from the intrusion

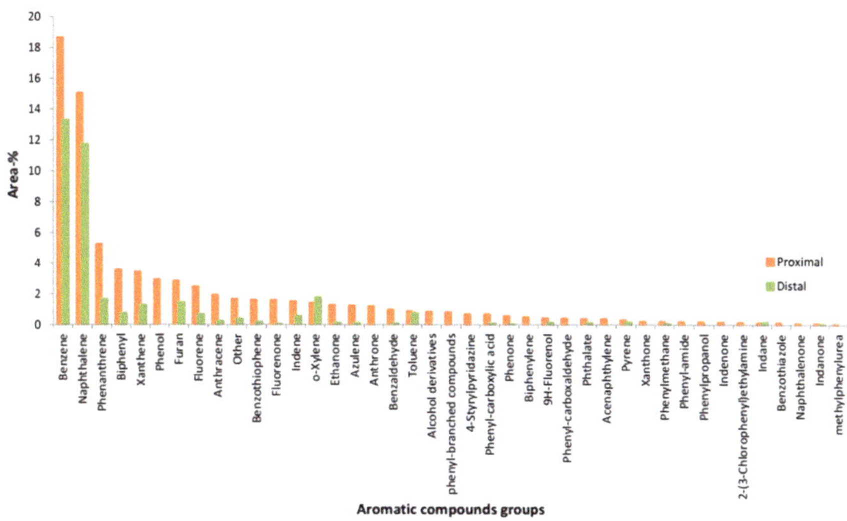

Fig. 4 Comparison of individual groups of aromatic organic compounds concentrations (in area-%) in relation to the distance from the intrusion

An aliphatic compound with the highest concentration evident in both coal samples is ethanol in which it contributes ~7.4 and ~36.6 area-% in proximal coal and distal coal, respectively. In addition, some of the aliphatic compounds such as thiourea, pentanone, pentenone, octane, hentriacontane, annulene and pentanediol diisobutyrate occur in very low concentrations (<1.2 area-%) in the proximal coal and are absent in the distal coal.

The aromatic compounds that make up these coal samples are more varied and regular than the aliphatic compounds. The organic compounds identified 41 subgroups composed of either entirely alkyl aromatic rings or a combination of both aromatic and aliphatic signatures. However, the organic compounds with both aromatic and aliphatic signatures (i.e. phenyl carboxylic acid) have the main functional group as aliphatic and the branches as aromatic (benzoic). Since the aromatic branches are significant, these compounds are classified into the 'aromatic compounds' group. Benzene, naphthalene, phenanthrene, biphenyl, xanthene, phenol, furan, fluorene and anthracene, inter alia, are the main aromatic compounds present in both coal samples in varying concentrations, especially in the proximal coal sample (Fig. 4). The proximal coal sample had the highest concentration for each aromatic compound compared to the distal coal with the exception being the concentration of o-xylene being more for the distal coal but with little variation. All these compounds consist of a variety of polycyclic and heterocyclic aromatic hydrocarbons.

Benzene as the main constituent of the aromatic compounds contributes ~18.6 area-% in the proximal coal and ~13.2 area-% in the distal coal. Benzene dominance is followed by naphthalene which composes ~15.1 and 11.7 area-% of proximal and distal coal surfaces, respectively. In Fig. 4, the proximal coal shows a gradual trend of decreasing concentrations in organic compounds from left to right.

4 Discussion

The sampled coal can be assumed to have gone through the same initial coalification processes prior to modification by the dolerite intrusions since it is in the same coal seam. The analyses of the organic chemistry on the proximal and distal coal samples to the dolerite intrusion show significant variations in the aliphatic and aromatic compounds under investigation. One can safely suggest that these variations may have been caused by the emplacement of the dolerite intrusions. In addition, Snyman and Barclay (1989) reported that the metamorphic effects of the dolerite intrusions affected the South African coal, while the coal was still at the lignite stage of coalification. The heat input and interaction of the intrusions with the coal may have influenced the coal properties and driving different reactions resulting in the observed variations. Thus, the coal in the study area reached the bituminous coalification stage succeeding the effects already caused by the heat input of the intrusions. As a rule of thumb, the contact metamorphism of the dolerite intrusion effect on coal extends at variable distances, generally at 0.2–6 times the thickness of the intrusive (Snyman and Barclay 1989). Thus, one can postulate that the 3-m-wide dolerite sill in the study area affected coal at maximum distance of 18 m, excluding the position of the distal coal at 20 m away from the intrusive.

The differences in detected chemical identification peaks in gas chromatograms between the proximal (1406 compounds) and distal (341 compounds) coal samples indicate that the heat input breaks down heavy macromolecules of the coal into smaller constituents. Consequently, this results in formation of new compounds that are present in proximal coal and absent in the distal coal (i.e. anthrone and thiourea). It is evident from the present aromatic compounds in the proximal coal (Fig. 4) that the more complex their aromatic rings becomes, the lower their concentrations in the coal samples but with gradual variations. Benzene and naphthalene are the aromatic compounds that dominate both coal samples, and this can be due to these organic compounds having the simplest and more stable aromatic structures and containing at most two aromatic rings. The results indicated that the aromatic compounds are abundant in the proximal coal constituting approximately triple the concentration of the aliphatic compounds, and this suggests that aromaticity decreases with increasing distance from the intrusion. This signature exhibited by the burnt coal, which is slightly of lower rank relative to the distal unburnt coal, is contrasting with the general idea that aromaticity of coal increases with increasing coal rank. However, this can be explained by Puttmann et al. (1991) who reported that lower rank coals are generally characterized by a mixture of trapped alkylated aromatic and saturated hydrocarbons. The aliphatic compounds constitute most of the distal coal, and therefore, 'aliphaticity' increases with increasing distance from the intrusion. An uncommon substantial concentration of ethanol together with other hydroxyls is evident in the distal coal. This abundance possibly resulted from hydrothermal alteration on the coal through fluids generated by the intrusion rather than primary coalification processes. Literature does not support the existence of such large amounts of hydroxyls. In fact, hydroxyl functional groups should get progressively expelled as coalification progresses or get remobilized as pyrolytic residue by devolatilization processes (Avila et al. 2014; Denisov and Metelitsa 1968; Emsbo-Mattingly and Stout 2011; McMurry 1996; Pone et al. 2007; Sasaki and Sugai 2011; Snyman and Barclay 1989).

The spontaneous combustion in coal is driven by the oxidation reactions involving the organic matter of the coal (Emsbo-Mattingly and Stout 2011) Aromatic compounds are stable and should not oxidize at the temperatures commonly accepted for spontaneous combustion (i.e. at 60 °C) (Emsbo-Mattingly and Stout 2011). Generally, the chemical reactions involving aromatic compounds in nature occur at temperatures above 200 °C with the presence of catalytic fluids enriched in Cu and Fe ions (Emsbo-Mattingly and Stout 2011). In contrast, aliphatic hydrocarbons will readily oxidize at low temperatures (McMurry 1996). Avila et al. (Avila et al. 2014) reported that highly reactive coals, the distal coal in this case, produce unstable solid oxygenated complexes (oxycoal), which are readily decomposed by the liberation of heat at low temperatures, and low reactive coals (i.e. proximal coal) produce very stable solid complexes that react more slowly and break down at much higher temperatures.

From the observed effects due to heat input by the intrusives, one can deduce that since aromaticity increases with increasing heat input upon the coal, spontaneous combustion or rather coal oxidation will be inhibited. The distal coal with substantial aliphatic compounds concentration is more prone to spontaneous combustion. However, if it occurs that both proximal and distal coal are subjected to oxidation (i.e. through ventilation), the more reactive distal coal will spontaneously combust earlier compared to the less reactive proximal coal which will take longer time to reach the self-heating temperature. As a result, the heat input by the dolerite intrusions has a positive impact on the organic matter of the proximal coal and a negative impact on the distal coal for the purpose of spontaneous combustion inhibition. However, this conclusion is only based on the chemical behaviour of the organic compounds among other intrinsic properties of coal that may behave differently to the metamorphic effects of the dolerite intrusions.

5 Conclusions

The spontaneous combustion of coal is a very complex process, and the processes still remain being not fully understood. The oxidation of coal is the principal mechanism leading to the initiation of the spontaneous combustion of coal. Among other factors affecting the spontaneous combustion, in the present study, the chemistry of the organic matter was utilized to evaluate the propensity of spontaneous combustion in the Highveld coalfields affected by dolerite intrusions. The functional groups detected from the chromatographic analyses are classified into aliphatic and aromatic compounds based on their reactivity when subjected to 19 oxidation processes, thus evaluating the concentration of these compounds within the proximal and distal coal. Aromatic compounds are stable and should not oxidize at the temperatures commonly accepted for spontaneous combustion, and in contrast, the aliphatic hydrocarbons will readily oxidize at low temperatures. The significant variations between the organic compounds' concentrations within the proximal coal and distal coal that underwent the same coalification processes suggest that the heat input by the dolerite intrusions had an effect on the coal. The proximal coal was subjected to maximum heat input and dominated by aromatic compounds. This observation thus suggests that the dolerite intrusion reduces the susceptibility of the spontaneous combustion in the proximal coal, however, based only on the chemical concentration of the organic compounds. The distal coal is more prone to spontaneously combust should the coal be exposed to ventilation.

References

Avila, C., Wu, T., Lester, E.: Estimating the spontaneous combustion potential of coals using thermogravimetric analysis. Energ. Fuels **28**(3), 1765–1773 (2014)

Denisov, E.T., Metelitsa, D.I.: Oxidation of Benzene. Russ. Chem. Rev. **37**(9), 656–665 (1968)

Emsbo-Mattingly, S.D., Stout, S.A.: Semi volatile hydrocarbon residues of coal and coal tar. In: Stracher, G.B., Sokol, E.V., Prakash, A. (eds.) Coal and Peat Fires: A Global Perspective, vol 1: Coal—Geology and Combustion, pp. 173–208. Elsevier (2011)

Finkelman, R.B.: Potential health impacts of burning coal beds and waste banks. Int. J. Coal Geol. **51**, 19–24 (2004)

Golab, A.: The impact of igneous intrusions on coal, cleat carbonate, and groundwater composition. PhD thesis, School of Geosciences, University of Wollongong (2003). Available from: http://ro.uow.edu.au/theses/164

McMurry, J.: Organic Chemistry, 4th edn. Brooks/ Cole Publishing Company. Pacific Grove, California (1996)

Pone, J.D.N., Hein, K.A., Stracher, G.B., Annegarn, H.J., Finkleman, R.B., Blake, D.R., Schroeder, P.: The spontaneous combustion of coal and its by-products in the Witbank and Sasolburg coalfields of South Africa. Int. J. Coal Geol. **72**(2), 124–140 (2007)

Puttmann, W., Steffens, K., Gobel, W.: Analysis of aromatic hydrocarbons in overburden from coal mines: assessment of the environmental impact. In: Peters, D.C. (ed.) Geology in Coal Resource Utilization, pp. 483–498. TechBooks, Fairfax, VA (1991)

Sasaki, K., Sugai, Y.: Equivalent Oxidation Exposure—Time for Low Temperature Spontaneous Combustion of Coal, Heat Analysis and Thermodynamic Effects. In Dr. Amimlu Ahsan (ed.). ISBN: 978–953-307-585-3 (2011)

Snyman, C.P., Barclay, J.: The coalification of South African coal. In: Lyons, P.C., Alpern, B (eds.) Coal: Classification, Coalification, Mineralogy, Trace-Element Chemistry, and Oil and Gas Potential. Int. J. Coal Geology, **13** 375–339 (1989)

Speight, J.G.: Handbook of coal analysis. In: Electronic Textbook. Wiley, New Jersey (2005)

Part II
Risk and Environmental Impact Assessment

Mud Inflow Risk Assessment in Block Caving Operation Based on AHP Comprehensive Method

A. Hekmat, A. Anani, F. Tapia and I. Navia

1 Introduction

The mining industry is almost considered as a highly potential risk industry due to the complexity and various sources of uncertainty. Among the mining methods, underground caving operations are naturally affected by mud and debris inflow; since, these methods are connected to surface or previous mining area with a broken subsidence zone, where is potential to accumulate water (Jacubec and Clayton 2012). Although mud rushes can have different origins, almost always four elements are required for a mud rush to occur: water, potential mud-forming material, a disturbance of mud, and discharge point (Butcher et al. 2000). At the mines that are likely to mud rush occurrence, a range of mitigating and controlling measures is adopted. Butcher et al. (Butcher et al. 2000) suggested a mud rush prevent approach based on three aspects: Keeping fine material far enough from the mining operations[1]; prohibiting water ingress into muck pile and proper definition of draw strategy to inhibit the discharge of hang ups, air blasts, and mud pockets. Heslop

[1]Site selection of waste dumps and tailing dams should be performed with the objective of eliminating the risk of material flow into underground operations.

A. Hekmat (✉) · F. Tapia
University of Concepcion, Concepcion, Chile
e-mail: ahekmat@udec.cl

F. Tapia
e-mail: franciscantapia@udec.cl

A. Anani
Pontifical Catholic University of Chile, Santiago, Chile
e-mail: angelina.anani@ing.puc.cl

I. Navia
El Teniente Copper Mine, Sewell, Chile
e-mail: isaacnaviamoreno@gmail.com

E. Widzyk-Capehart et al. (eds.), *Proceedings of the 18th Symposium on Environmental Issues and Waste Management in Energy and Mineral Production*,
https://doi.org/10.1007/978-3-319-99903-6_5

Table 2 Random consistency (RC) index [n = size of the reciprocal matrix]

n	1	2	3	4	5	6	7	8	9	10
RC	0	0	0.58	0.9	1.12	1.24	1.32	1.41	1.45	1.49

(2) Summed each column vector of $\bar{A}$:

$$\overline{w_i} = \sum_{j=1}^{n} \bar{a}_{ij} \quad (i = 1, 2, \ldots, n) \tag{3}$$

(3) Normalized each vector of $\bar{W} = (\bar{w}_1, \bar{w}_2, \ldots, \bar{w}_n)$:

$$w_i = \frac{\bar{w}_i}{\sum_{i=1}^{n} \bar{w}_i} \tag{4}$$

Saaty (1990) suggested using the maximum eigenvalue method to determine the judgment matrix as:

$$A \times W = \lambda_{\max} \times W \tag{5}$$

where $\lambda_{\max}$ is the maximum eigenvalue of the matrix A.

For a reliable comparison, it is important to note that the inconsistency of the comparison matrix A must be less than 10%. The consistency is defined by relation between the entries of A: $a_{ij} \times a_{jk} = a_{ik}$. According to Saaty (1990), the consistency of judgments can also be evaluated using the Eq. (6):

$$\text{Consistency ratio} = \text{CR} = \frac{\text{CI}}{\text{RC}} \tag{6}$$

CI is the consistency index and is defined as:

$$\text{CI} = \frac{\lambda_{\max} - n}{n - 1} \tag{7}$$

The random consistency index (RC in Eq. 6) can be obtained from Table 2. Since the column(s) of any 1×1 or 2×2 comparison matrices are dependent, RC is assumed to be 0. This means division by zero in Eq. (6) and causes CR to tend toward infinity; that is, matrices of sizes 1 and 2 are always consistent (Aminbakhsh et al. 2013).

3 Case Study

This study aimed to evaluate the mud entrance risk in one sector of the world's largest copper–molybdenum undergrounds mine, El Teniente which is owned by Codelco, Chile. Respecting the reserve size, El Teniente is classified as the sixth largest copper deposit in the world. This mine is located at the 70 km south-southeast of Santiago in the Andes mountain range, Chile. The mine uses block and panel caving methods to extract ore with the daily production rate of about 140,000 tonne and the mean grade of 0.86% copper. Mining is carried out at different levels around a non-mineralized formation called the Braden Pipe that houses mining infrastructure of each level. The current mine level contains six mining blocks around the Braden Pipe at different elevations including the Esmeralda, Reservas Norte, Diablo Regimiento, and Pipa Norte mining blocks (Fig. 1).

Located in Ands mountain range, the main source of water inflow is surface water, especially winter snow melting during spring (Ferrada 2011). Additionally, underground water and the water which is used for hydraulic fracturing are the other sources of water appearance in the operating area. Drainage system is piping water along the main conveyor belt to the surface. Among the 14 operating sectors in 2011, five sectors have been faced with the problem of mud entrance of which Diablo Regimiento is one of them.

A statistical study of humidity, size distribution, and production characteristics of 227 draw-points at Diablo Regimiento (D.R) sector was accomplished in this research. D.R sector is in the south part of the deposit. This sector is divided into five planning phases and is placed under three previous mined sectors, of which one of those (Regimiento) had been closed due to the existence of mud. Mud entrance study of this sector showed that the initial entry of mud into D.R was due to the connection of this sector to the upper mined sectors. Once mudflows into the caved column, it immigrated laterally into other parts of D.R based on draw control and mine planning strategies.

4 Mud Inflow Risk Affecting Parameters

To evaluate the risk of mud inflow at D.R, a risk-based hierarchy was constructed considering the potential considering the potential risk items threatening the safety of mining operation in terms of mud inflow into the extraction points. The influence of each parameter in mud inflow was studied based on the historical database of the mine. The hierarchy was constructed comprising three criteria, each of which was then divided into different sub-criteria (Fig. 2). The results of statistical analyses of all these elements were used to make the pair-wise comparison among them.

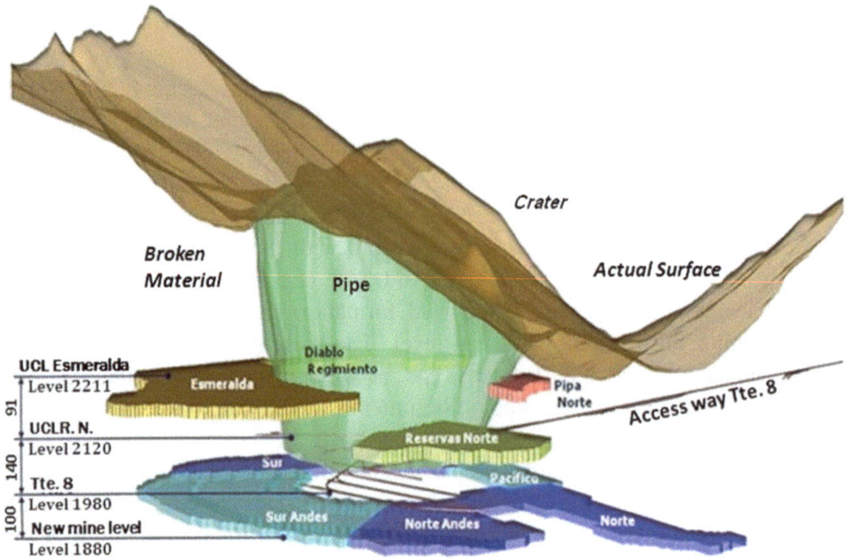

Fig. 1 Location of different sectors at El Teniente mine (CODELCO 2009)

4.1 *Fine Material*

According to Butcher (2005), the percentage of fine material in extraction point is one of the four required factors to provoke mud rush in cave mining. Considering the caving circumstances, generation of fine material in extraction points is an inevitable process, due to extended friction between fragments, secondary fragmentation, geological conditions, and fine migration. The common size classification in the D.R sector is presented in Table 3. Generally, the material with the size less than 5 cm is considered as fine material. Historical review of several underground mines shows that the size less than 5 cm and/or less than 25 cm are the most perilous sizes in mud generation.

4.2 *Water Content*

The other important parameter to produce mud, which was identified by Butcher (2005), is the existence of water. Water has adverse influence on the mud entrance development. Without water, there is no risk of mud rush, even if all of the other criteria are existing (Jacubec and Clayton 2012). Likewise, in the absence of fine material, the water will flow through the bulk material without any risk of mud rush occurrence. Thus, these two factors should be measured regularly at each extraction

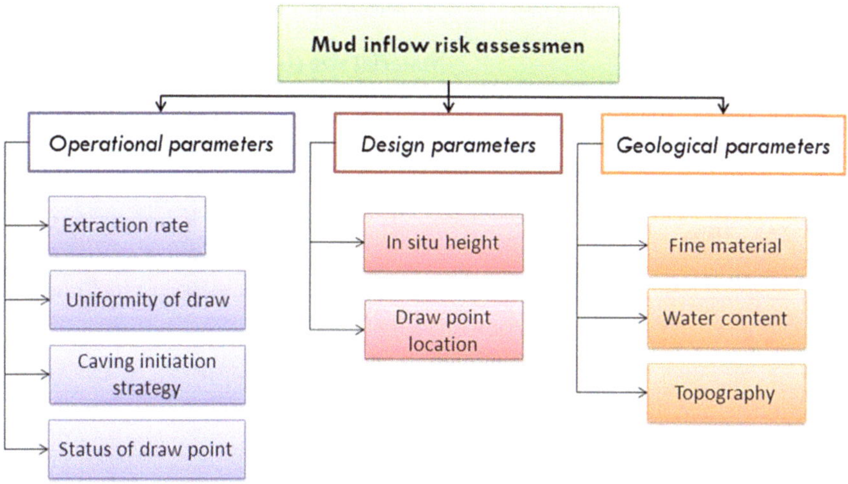

Fig. 2 Hierarchy of mud inflow risk affecting parameters

Table 3 Size distribution classification of different status of draw-points (with mud and dry)

Size classification	No. of data		Average percentage		Standard deviation	
	With mud	Dry	With mud	Dry	With mud	Dry
Less than 5 cm	68	20182	60.88	30.51	21.94	24.95
Between 5 and 25 cm			26.91	33.77	12.49	17.41
Between 25 and 50 cm			9.63	21.75	12.20	16.70
Between 50 and 100 cm			2.43	9.49	6.94	15.37
More than 100 cm			0.15	4.04	1.21	12.64

Table 4 Humidity classification in El Teniente mine

Class code	Qualitative expression	Moisture percentage
0	Dry	0%
1	Low humidity	Less than 4%
2	Humid	4–7%
3	Mud incipient	7–10%
4	Mud	More than 10%
A	Coarse material and water flow	

Table 5 Wet muck classification matrix at El Teniente copper mine

Humidity content	Material size (G) ≤ 25 cm		
	G < 30% (Coarse material)	30% ≤ G < 70%	G ≥ 70% (Fine
< 4%	Green	Green	Green
4% - 7%	Yellow	Yellow	Yellow
7% - 10%	Yellow	Yellow	Red
≥ 10%	Red	Red	Red

Green	**Normal condition**
Yellow	**Mud observation**
Red	**Critical risk**

point. In El Teniente mine, draw-points are classified into five groups regarding the humidity percentage (Table 4).

The last two classes in Table 4 (classes 3 and 4) are defined as the highest risk of mud entrance into the operation area. According to Tables 3 and 4, a risk matrix was developed by the risk evaluation center to manage the mud rush risk at the mine (Table 5). Based on Table 5, the draw-points with a high risk of mud entrance are closed to prevent the hazards of mud rush.

4.3 Topography

Surface crater monitoring is one of the preventive measurements in the operational mud rush risk assessment. In the case of caving operation, the internal and external mud rushes are likely to occur because they connect to the surface with a caved zone which provides a potential point of entry for water and mud. On the positive side, when mines go dipper, correlation between topography and external mud rush inflows diminished. However, the connection of caved column with previous mined sectors would be influential. Figure 3 shows the surface topography and the mud advance zone in D.R. It is obvious that mudflow appears in draw-points where located beneath the subsidence zone with the less distance from the potential area of water accumulation in the surface.

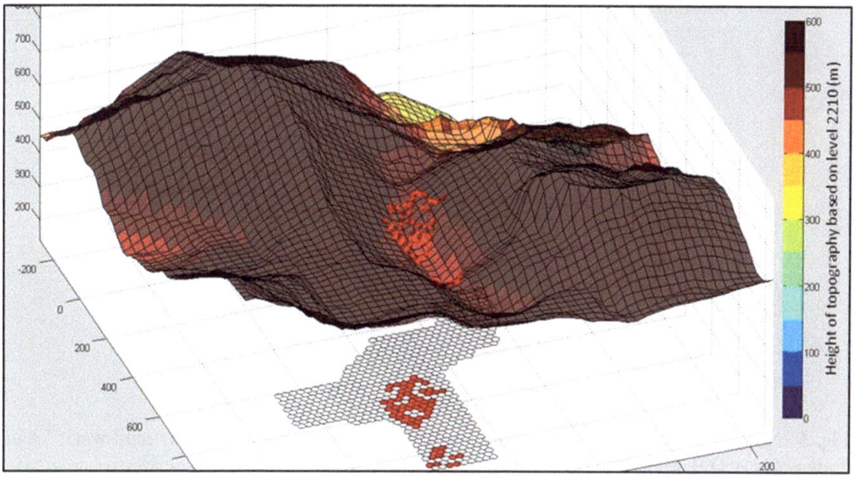

Fig. 3 Surface topography and mud advance zone in D.R

4.4 In Situ Height

Analyzing different conditions of the first mud observation in a draw-point shows the significant influence of extraction strategy and height of broken column on mud entrance into the working area. Figure 4 illustrates the mud presence date and extraction height of draw-points in D.R. from 2009 to 2013. It is obvious in this chart that the mud is discovered at the height of more than 100 m which is the height of in situ column, between D.R sector and its upper sector, Regimienro (R). Thus, it can be concluded that the risk of mud entrance when extraction height is less than the in situ height is almost zero.

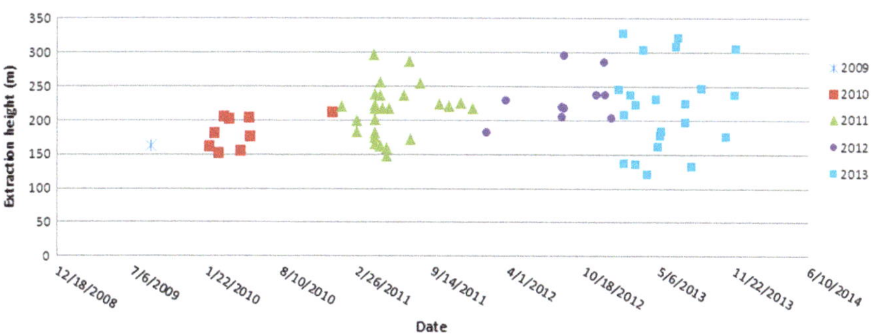

Fig. 4 Extraction height of the first mud appearance in draw-points at D.R

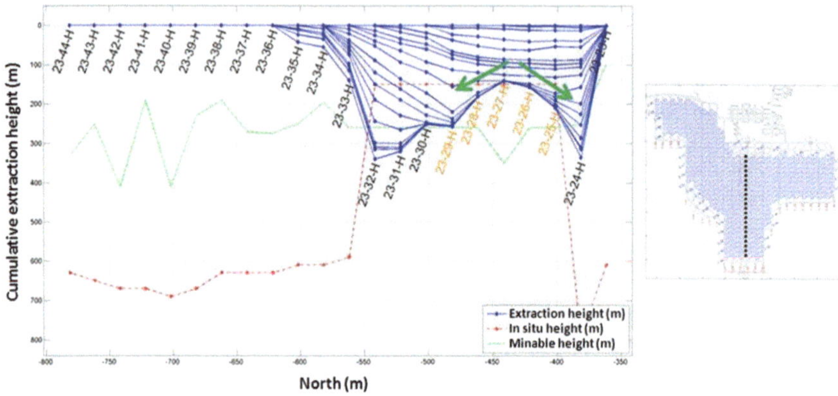

Fig. 5 North–south section of extraction height profiles along the first mud "mud-water" status draw-point at D.R from 2005 to 2013

4.5 *Uniformity of Draw*

To characterize the mudflow behavior, it is required to consider drawing manner of each draw-point over time. Based on the extraction height profile, it is possible to qualitatively evaluate the uniformity of draw by determining how a profile is equidistance comparing to its previous profile at a certain time.

Figure 5 shows the extraction height profile along north–south sections of the first "mud-water" status draw-point at D.R during 2005–2013. This draw-point (with the ID number of 23–27-H) has been closed on March 2009, due to the high risk of mud rush. In this figure it is obvious from the annually extraction heights that a specific draw control discipline were applied a no uniform profile with the objective of providing arch failure to insure progressive caving. However, implementing this type of drawing strategy yields preferential flow of mud, which caused mud inflow from the zones with maximum profile height toward the lowest points in the profile (Fig. 5).

4.6 *Caving Initiation Strategy*

Production planning of D.R was carried out with the objective of creating a continuous propagation of caving by generating the active volume in the center of the sector. Therefore, extraction was started from the specific draw-points in the center of the first production phase of the sector, and the broken "arch" reaches the former sector as shown in Fig. 6. The historical data analyses showed that mud entrance appeared in the center of the sector when the active volume reached the overlying mine sector.

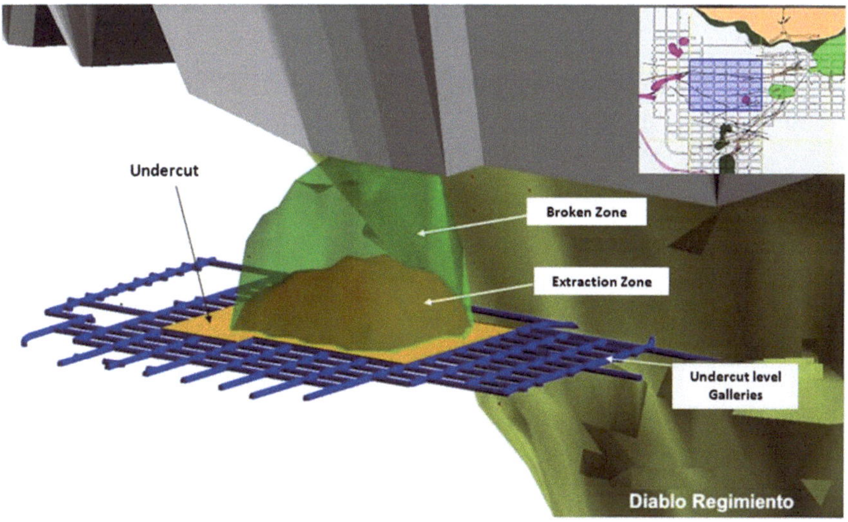

Fig. 6 Status of extraction zone at the time of caved zone's connection to the former mining sector

4.7 Status of Extraction Point

According to Table 4, draw-points with the most probability of mud rush occurrence are classified as "mud-water status" and have been closed permanently to inhibit the entrance of mud to the working area. Preventing mud entrance, to the working area, results in mud distribution into the adjacent extraction points. Under those circumstances, the neighborhood draw-points would have the hazards of lateral immigration of mud. The extraction rate of these points should be controlled to avoid mud migration into the operative extraction points. For this reason, these draw-points are allocated as "limited status." Furthermore, some draw-points are used as barrier to control the entry of mud. These draw-points are labeled as "barrier status." In this status, the content of moisture and fine material are not essentiality critical. However, they are considered as high-risk draw-points regarding the flow direction of mud.

Owing to the dynamic condition of caving procedure, the status of extraction point changes continuously during extraction. Figure 7 illustrates mud distribution to different draw-points during five-year operation. It is obvious in Fig. 7 that the status of draw-points at the vicinity of "mud-water" draw-points changed and mud distribute into other points and change their status during operating years.

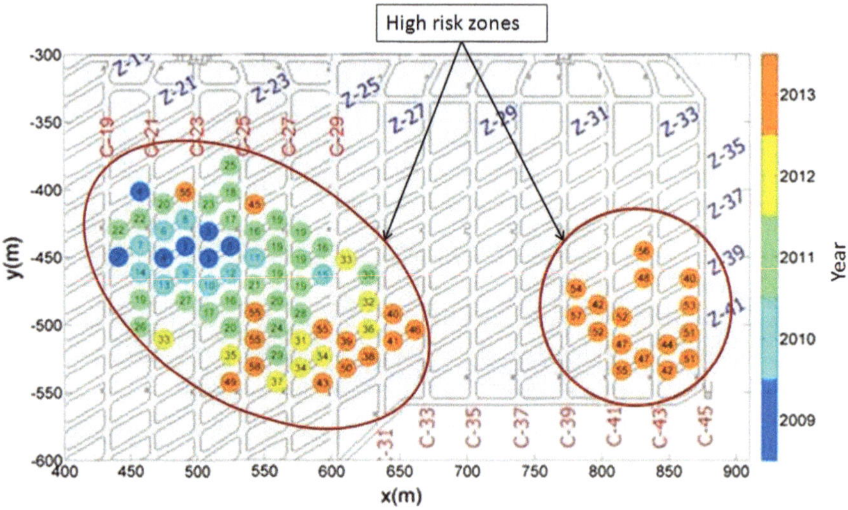

Fig. 7 Mud appearance sequence in D.R from 2009 to 2013

4.8 Extraction Rate

Controlling the tonnage drawn from individual draw-points could avoid creating the conditions that could lead to mud rushes (Laubscher 2000). Analyzing the extraction rate of different draw-points showed that the frequency distribution function of extraction rate at draw-points with late mud entrance was almost uniform. Moreover, the cumulative distribution function (CDF) of extraction rate for different draw-points revealed that the extraction rate of early mud entrance draw-points was almost less than that in the late mud entrance draw-points. According to this study, increasing the extraction rate in a uniform manner will significantly decrease the mud inflow risk.

4.9 Draw-Point Location

The historical study of mud entrance into the draw-points at D.R sector shows that these draw-points are mostly located beneath the former mining sectors with mud occurrence. Figure 8 shows a plan view of D.R in which the position of muddy draw-points is compared with the muddy draw-points in its upper sector (Regimiento-Pink bullets). It is obvious in Fig. 8 that the draw-points which are placed under the muddy former levels have a higher risk of mud entrance than others.

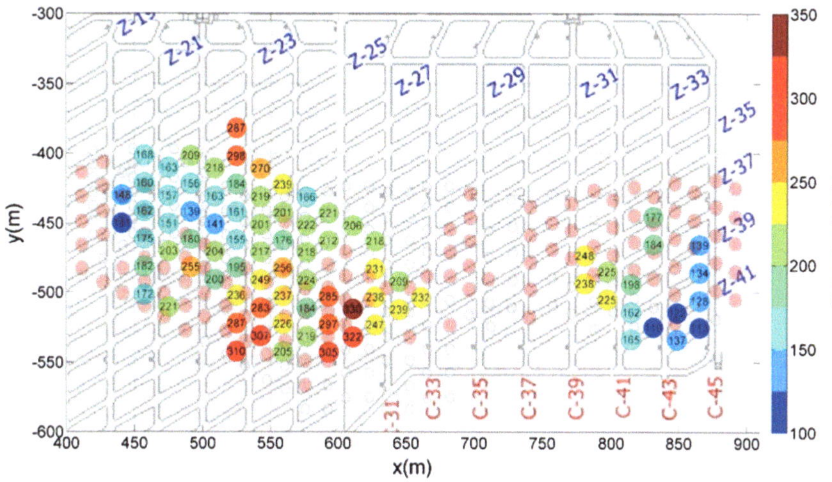

Fig. 8 Location of muddy draw-points in D.R and Regimiento sectors

	Operational Parameters	Design aspects	Geological parameters
Operational Parameters	1	2	4
Design aspects	0.50	1	2
Geological parameters	0.25	0.50	1

	Operational Parameters	Design aspects	Geological parameters	Weighting
Operational Parameters	0.57	0.57	0.57	0.57
Design aspects	0.29	0.29	0.29	0.29
Geological parameters	0.14	0.14	0.14	0.14

Operational parameters

	E.R	U.D	C.I.S	DP.S
E.R	1	9	3.62	4.79
U.D	0.11	1	0.41	0.54
C.I.S	0.28	2.47	1	1.32
DP.S	0.21	1.87	0.76	1

	E.R	U.D	C.I.S	DP.S	Weighting	Overall weight
E.R	0.63	0.63	0.63	0.63	0.63	0.36
U.D	0.07	0.07	0.07	0.07	0.07	0.04
C.I.S	0.17	0.17	0.17	0.17	0.17	0.10
DP.S	0.13	0.13	0.13	0.13	0.13	0.07

Design parameters

	IS.H	DP.L
IS.H	1	0.43
DP.L	2.30	1

	IS.H	DP.L	Weighting	Overall weight
IS.H	0.30	0.30	0.30	0.09
DP.L	0.70	0.70	0.70	0.20

Geological parameters

	F.M	W.C	T
F.M	1	0.97	6.43
W.C	1.03	1	6.63
T	0.16	0.15	1

	F.M	W.C	T	Weighting	Overall weight
F.M	0.46	0.46	0.46	0.46	0.07
W.C	0.47	0.47	0.47	0.47	0.07
T	0.07	0.07	0.07	0.07	0.01

Fig. 9 Pair-wise comparison matrices

5 AHP in Mud Inflow Risk Assessment

The AHP framework was illustrated in Fig. 2 which arranged the most influenced parameters on mud inflow into three main categories. The objective of this framework is to provide a decision tool to determine the adequate investments for

Table 6 Odd ratio of mud inflow influenced parameters

Criteria	Sub-criteria	Odd ratio
Operational parameters	Extraction rate	13.4
	Uniformity of draw	1.5
	Caving initiation strategy	3.7
	Status of draw-point	2.8
Design parameters	In situ height	1
	Draw-point location	2.3
Geological parameters	Fine material	6.3
	Water content	6.5
	Topography	0.98

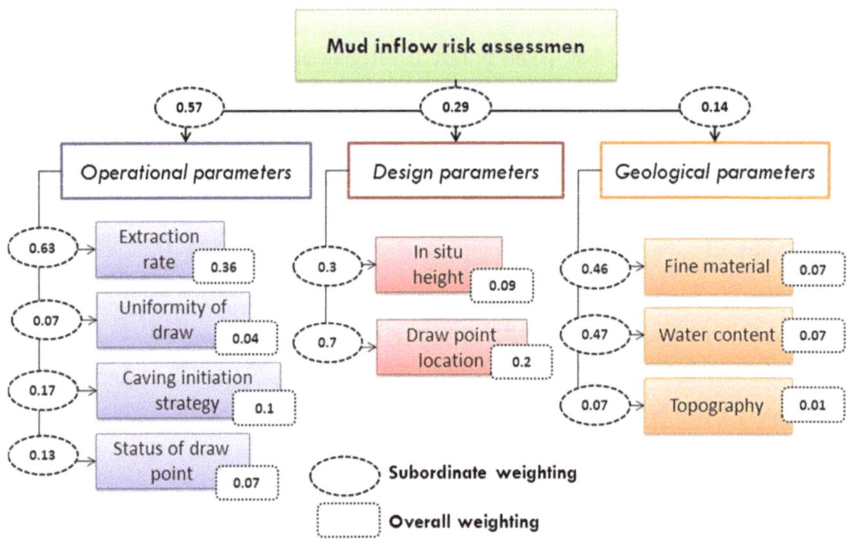

Fig. 10 Priority weights of mud inflow risk influenced parameters

mud inflow prevention. Based on the hierarchy shown in Fig. 2, four reciprocal matrices were constructed to making the pair-wise comparisons among the elements of the hierarchy (Fig. 9).

The first pair-wise comparison in Fig. 9 was made among the parameters on the top level of the hierarchy (criteria). In this comparison, the highest weight was assigned to the operational parameters since these parameters are the most controllable ones, while the lowest weight was allocated to the geological parameters since they are less likely to be able to change. The odd ratio (OR) of the logistic regression between each sub-criterion and mud inflow was considered to make the pair-wise comparison of these elements (Table 6). The odd ratio is one way to quantify how strongly the presence or absences of elements in the inflow risk hierarchy are associated with the mud inflow into the working area. If the OR is greater than one, then having that element is considered to be associated with having mud inflow.

In order to insure the consistency of the judgments in all the reciprocal matrices, consistency ratios (CR) were calculated using the largest eigenvalues of the eigenvectors (Eqs. 5–7). Figure 10 shows the normalization weights for each of the elements in the hierarchy as well as the overall weightings. As it is illustrated in Fig. 10, the overall prioritization revealed that the "extraction rate" is the item with the most significant impact. Hence, to prevent mudflow risk, extraction rate requires the most significant investment and consideration among the other elements.

6 Conclusion

Project safety risk assessment is the fundamental component of the project management since the efficiency of mining projects is highly influenced by safety problems such as mud rush, rock burst, air blast. In this paper, a framework was proposed to compare the most effective parameters on mud inflow into the extraction points at El Teniente copper mine, Chile. This method provides a tool for the mine managers to define the priority of influenced items to create a rational budget for mud rush prevention during the planning and feasibility study of block caving mining projects. The proposed framework was applied to a real mining project to illustrate how it can guide the decision makers in mud inflow risk assessment.

According to the overall weight of different elements, it is concluded that extraction rate has the first priority to be taken into consideration to avoid the risk of mud inflow occurrence. The second order belongs to the draw-points which are located beneath the former mining sectors with mud existence. The third preference of caving initiation strategy shows that the mud rush risk should be considered when applying an undercutting strategy. Even though the fine material and water content have a significant effect on mud inflow, these parameters are located in the fifth rank, since the geological parameters had the lowest subordinate weights. It is important to realize that even by applying drainage systems and draw strategies, the persistence of water and fine material is unavoidable in mining operation. However, controlling the extraction rate and uniform draw would decrease the probability of fine percolation and water inflow.

Acknowledgements The authors would like to acknowledge the financial support of the Chilean Government through the CORFO project 12IDL2-15145.

References

Aminbakhsh, S., Gunduz, M., Sonmez, R.: Safety risk assessment using analytic hierarchy process (AHP) during planning and budgeting of construction project. J. Saf. Res. **46**, 99–105 (2013)

Bascetin, A.: Mining technology: Transaction of the Institution of mining and metallurgy. In: Technical note: an application of the analytic hierarchy process in equipment selection at Orhanel open pit copper mine, pp. 192–199 (2004)

Bhushan, N., Rai, K.: Strategic decision making—applying the analytic hierarchy process. Springer press (2004)

Butcher, R., Joughin, W., Stacey, T.R.: The safety in mines research advisory committee (SIMIAC). In: A booklet method of combating mudrushes in diamond and base metal mines, pp. 1–35 (2000)

Butcher, R., Stacey, T.R., Joughin, W.C.: Mud rushes and methods of combating them. J. South Afr. Inst. Min. Metall. **105**, 817–824 (2005)

Dagdeviren, M., Yüksel, I.: Developing a fuzzy analytic hierarchy process (AHP) model for behavior-based safety management. Inform. Sci. **178**, 1717–1733 (2008)

Ferrada, M.: Gravity flow under moisture conditions—control and management of drawpoint mudflow. In: Proceedings of the 35th APCOM symposium, Wollongong, NSW (2011)

Hekmat, A., Osanloo, M., Akbarpour Shirazi, M.: A new approach for selection of waste dump site in open pit mines. J. Min. Technol. **117**, 24–31 (2008)

Heslop, T.G.: Block caving—controllable risk and fatal flows. In: Proceedings of MassMin, Brisbane, Australia, pp. 437–454 (2000)

Jacubec, J., Clayton, R., Guest A.R.: Mudrush risk evaluation. In: Proceedings of MassMin, Greater Sudbury, Ontario, Canada (2012)

Kousalya, P., Reddy, G.M., Supraja, S., Prasad, V.S.: Analytical hierarchy process approach—an application of engineering education. Math. Aeterna 2 **10**, 861–878 (2012)

Laubscher, D.: International caving study. A practical manual on block caving (2000)

Lee, M.G.: Information security risk analysis methods and risk trends: AHP and fuzzy comprehensive method. Int. J. Comput. Sci. Inform. Technol. **6**, 29–45 (2014)

Mustafa, M.A., Al-Bahar, J.F.: Project risk assessment using the analytic hierarchy process. IEEE Trans. Eng. Manag. **38**, 46–52 (1991)

Saaty, T.L.: The Analytic Hierarchy Process. MCGrow-Hill, New York (1980)

Saaty, T.L.: Decision making with the AHP: why is the principal eigenvector necessary. Eur. J. Oper. Res. **145**, 85–91 (2003)

Saaty, T.L.: How to make a decision—the analytic hierarchy process. Eur. J. Oper. Res. **48**, 9–26 (1990)

Yurdakul, M.: AHP as a strategic decision-making tool to justify machine tool selection. J. Mater. Process. Technol. **146**, 365–376 (2004)

The Impact of Sand Mining on the Fluvial Environment: Case Study of Nzhelele River in Limpopo Province, South Africa

F. Sengani and T. Zvarivadza

1 Introduction

Sand is a commodity mostly used in construction industry, especially roads construction, building sites, brick making and reclamation of disturbed environments. Padamalal and Maya, 2014, outlined that sand mining has economic importance, and for centuries, humans have been enjoying the natural benefits provided by sand mining from rivers without understanding much on how the river ecosystem functions and maintains its vitality. Among the various types of human interventions in river ecosystems, sand mining is the most disastrous activity as it threatens the very existence of these systems. Due to the large demand for sand in South Africa, illegal sand mining is increasing at a high rate. Although illegal sand mining occurred in small-scale areas, the rate of sand extraction exceeds the natural sediment yield of the river system throughout the country (Chevallier 2014). Carnie (2015) discussed that there is evidence of coastal erosion, beaches and facilities washed away by a storm in Durban due to unsustainable illegal sand mining. This can lead to an idea that sand mining has large economic, social, ecological and environmental consequences for South Africa and local resource users.

Categorized the impacts of sand mining into three classifications including:

- physical impacts which are results of sand mining from streambed causing alteration of channel slope and changes in channel morphology;

F. Sengani (✉) · T. Zvarivadza
School of Mining Engineering, The University of the Witwatersrand, Johannesburg, South Africa
e-mail: fhatugeorge@gmail.com

T. Zvarivadza
e-mail: tawanda.zvarivadza@wits.ac.za

E. Widzyk-Capehart et al. (eds.), *Proceedings of the 18th Symposium on Environmental Issues and Waste Management in Energy and Mineral Production*,
https://doi.org/10.1007/978-3-319-99903-6_6

- water quality impacts which are caused by sand mining and dredging activities resulting in compromising of the water quality for downstream users and increase treatment costs;
- ecological impacts such as loss of habitats and species disturbance, removal of channel substrate and suspension of sediments, which are results of sand mining.

1.1 Problem Statement

Nzhelele villages are expanding at an alarming rate, and such expansion led to growth in infrastructure, residential areas and construction of new roads (Bumby 2000; Carnie 2015). The increase in demand for construction in the study area led to an increase in rate of sand mining which then results in several problems such as deterioration of the quality of river water, river bank erosion, floodplain deforestation and river bed degradation (Brandl 2002; Chevallier 2015; Dallas and Day 1993; Duffy and Phyllis 2007). Disproportionate mining of sand jeopardizes the health of river in particular and the quality of the environment in general. Sand mining is incompatible with the natural environment because of its negative environmental impacts. The removal of topsoil to expose the underlying quality material degrades the aesthetic beauty of the natural environment thereby forming unpleasant landscapes. The abandoned and non-rehabilitated sites lose value and therefore change the land use of an area (Chevallier 2014; Garton 2015; Gwimbi and Dirwai 2003; Haupt and Sami 2006).

Therefore, a detail research project on the evaluation of the environmental impacts caused by sand mining was conducted through the use of geological principles, geotechnical principles, geographical information system (GIS) and remote sensing to assess the environmental impacts of sand mining along Nzhelele River Valley.

2 Methodology

The river was divided into three sections (upstream, middle stream and downstream) for the purpose of data collection. Field measurements of water (temperature, pH, EC, TDS and turbidity) and dust were performed using standard methods of analysis as prescribed by South African National Standards. Owing to that soil and water samples were also collected with the purpose of undertaking laboratory analysis such as sieve analysis (grains distribution) and turbidity, total dissolved solids (TDS), pH, temperature and electric conductivity were analysed using Jenway pH and conductivity metre. In order to validate the results of the study and identification of the historical information about the study area, GIS methods [multi-temporal images (Landsat ETM + Images) and normalized difference vegetation index (NDVI)] were applied.

3 Results of the Study

In order to analyse the impacts of sand mining on the water quality, the river was divided into three sections (upstream, middle stream and the downstream). The downstream section of the river was assumed to be unaffected by sand mining and was used as a reference section. Sand mining was taking place along upstream and middle-stream sections during the investigation. However, most portions of the upstream and middle-stream section were noted to be previously exposed to sand mining activity and mining was in progress. Water samples were collected from three sections for a period of six months, with the purpose of analysing physical water quality parameters (turbidity, pH, temperature, total dissolved solids and electrical conductivity). Temperature and pH of water samples were analysed in the field and also in the laboratory, and the average values were taken. All the analyses were compared to the required standard of water quality speculated by South African Department of Water Affairs (1996 and 2004).

3.1 Temperature of Water at Nzhelele River

Previous study by Dallas and Day (1993) reported that water temperature plays a significant role in bodies by affecting the rate of chemical reactions and metabolic rate of organism. The study shows that higher temperature reduces the solubility of dissolved oxygen in the water, decreasing its concentration and its availability to aquatic organisms (Dallas and Day 1993; Madise 2013; World Health Organization 1996). Based on the results of the study, it was noted that the water temperature measured during the investigation was found to range from 24.1 to 35 °C (see Fig. 1). The water temperature was found to be increasing gradually in both upstream and middle stream with constant water temperature along the downstream section. However, due to sand mining activities which were taking place along the upstream and middle-stream section, it was then concluded that the water bodies were exposed to sunlight due to the extensive removal of vegetation parts of the river to heat. Based on the SADWAF, water temperature above 20 °C is not suitable for the aquatic organism, due to that both upstream and middle-stream sections were affected and could not be able to support the growth and survival of aquatic organism.

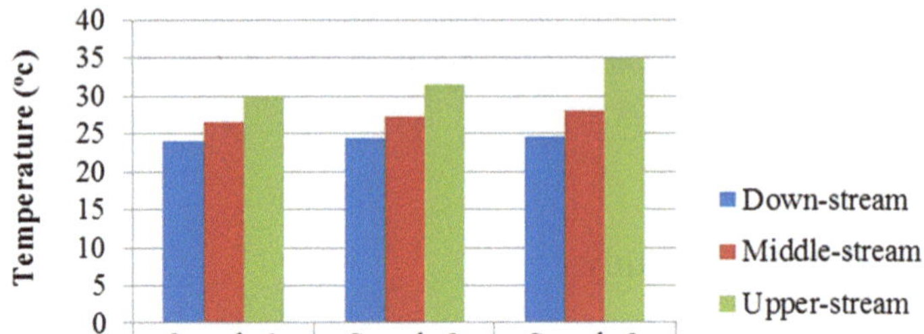

	Sample 1	Sample 2	Sample 3
Down-stream	24,1	24,3	24,5
Middle-stream	26,5	27,3	28
Upper-stream	30	31,5	35

Fig. 1 Temperature analysis of river water at Nzhelele River

3.2 *Electrical Conductivity of the Water*

The electrical conductivity of the water is a measure of the capability of a solution such as water in rivers to pass an electric current (Bruvold and Ongerth 1969; South Africa 1996). The recommended maximum limit for electrical conductivity for domestic water quality is 700 us/cm (South Africa 1996, 2004; World Health Organization 1996). The electrical conductivity of the study area was found to range from 70 to 214 us/cm. Although the water temperature of the upstream and middle stream was found to be higher than the recommended standard, the electrical conductivity along the stream was found to be within the required limit for domestic water as per the World Health Organization (1996) recommendations (see Fig. 2). Although the electrical conductivity of the water was found to be within the limit, it was also noted that the rapid increase in electrical conductivity within the upstream and middle stream will eventually exceed the limit if sand mining progresses at high rate within the sections. This significant increase in electrical conductivity of water along the upstream and middle stream can be an indication of pollutant discharges that have entered the river as a result of extensive sand mining.

Current experiment by Duffy and Phyllis (2007) has outlined that the total dissolved solids are defined as the measurement of inorganic salts, organic matter and other dissolved materials in water. Further analysis showed that the total dissolved solids values always increased with the increase in electrical conductivity of the water. Similar to electrical conductivity results, higher total dissolved solids values (151.6–197.7 mg/l) were measured along the upstream section followed by the middle-stream section and with low total dissolved solids within the lower-stream section (see Fig. 3). It was then concluded that the total dissolved solids within the river water were within the required limit, but rapid increase in sand mining has the ability to increase the total dissolved solids in the river.

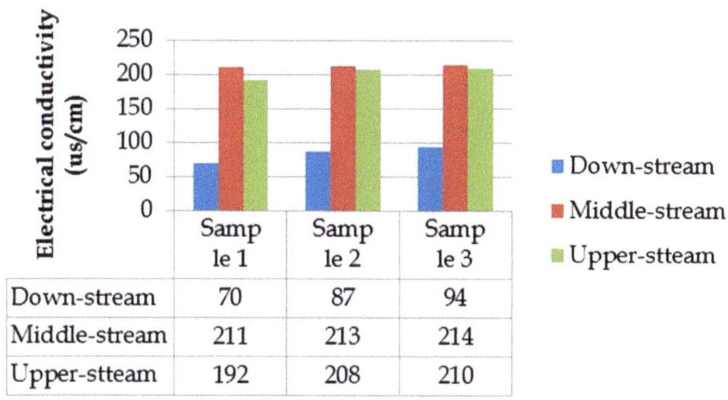

	Samp le 1	Samp le 2	Samp le 3
Down-stream	70	87	94
Middle-stream	211	213	214
Upper-stteam	192	208	210

Fig. 2 Electrical conductivity of Nzhelele River Valley

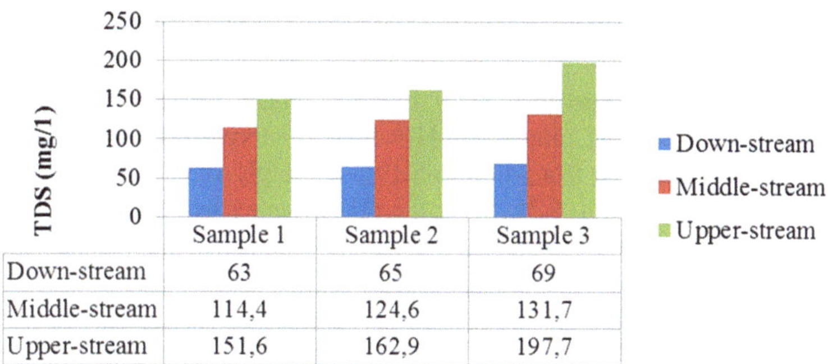

	Sample 1	Sample 2	Sample 3
Down-stream	63	65	69
Middle-stream	114,4	124,6	131,7
Upper-stream	151,6	162,9	197,7

Fig. 3 Total dissolved solids for river water

3.3 Turbidity of the Water at Nzhelele River Valley

Turbidity is an important indicator of the suspended sediments' quantity in water, and this has many negative impacts on aquatic life. The results of the study show that high turbidity was measured along the upstream and the middle stream, and very low turbidity was measured at the downstream. Owing that it was then concluded that riverbank erosion that occurred due to extensive sand mining has influenced the increase in turbidity level along the two sections of the river, besides that vegetation removal has also found to contribute towards the increase in turbidity due to factor that vegetation retain soil particles together with their roots (see Fig. 4).

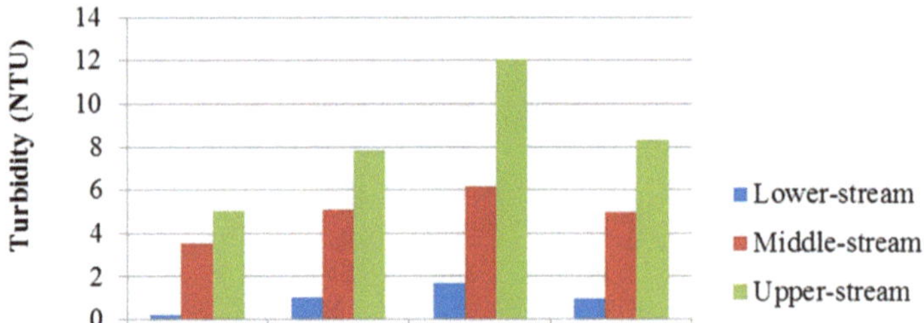

	Sample 1	Sample 2	Sample 3	Average
Lower-stream	0,2	1,02	1,73	0,98
Middle-stream	3,54	5,09	6,15	4,98
Upper-stream	5,01	7,8	12,02	8,28

Fig. 4 Turbidity of river water at Nzhelele River

3.4 pH Value of the Water at Nzhelele River Valley

It is important to measure the pH of the river water because different aquatic organisms prefer different ranges of pH values to survive. The pH values of the midstream section ranged from 4.62 to 5.19 and the pH values of the downstream section ranged from 5.20 to 6.35. Since the pH values of the midstream section and the downstream section are less than 7, water within the two sections was considered to be acidic and harmful to the immature fishes and insects that are available at Letaba River. The pH values of the upstream found to range from 7.23 to 7.56. However, the water pH at upstream was considered to be neutral to early basic in most of the organisms that can be able to survive (see Fig. 5).

3.5 Air Quality

Air quality measurements were conducted using gravimetric dust pumps. It was noted that the concentration of airborne dust exceeded 2 mg/m^3 during

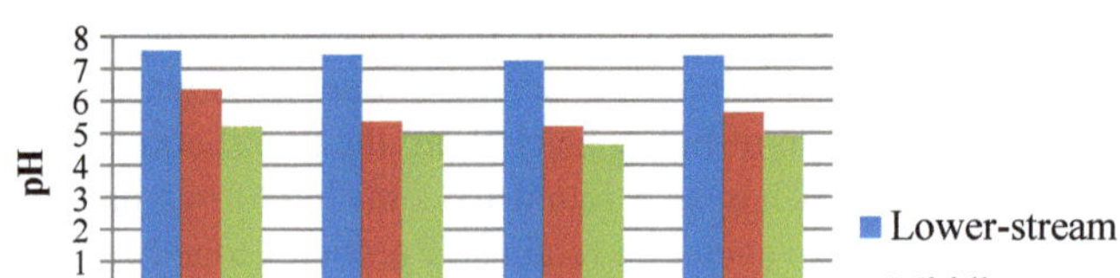

	Sample 1	Sample 2	Sample 3	Average
Lower-stream	7,56	7,42	7,23	7,4
Middle-stream	6,35	5,34	5,2	5,63
Upper-stream	5,19	4,95	4,62	4,92

Fig. 5 pH values of Nzhelele River

Table 1 Gravimetric test results (suspended dust concentration in mg/m^3)

Day 1		Day 2	Day 3	Day 4	Day 5
Downstream	1,4	1	1.2	1.3	1.2
Middle stream	1.6	2.6	4.3	4	3.2
Upstream	1.9	3.3	5	4.5	3.4

transportation and loading of sand and more effective during sunny days. It was also found that dust level was ranged from 1.9 to 3.4 mg/m^3 on downstream and 1.6 to 3.2 mg/m^3, upstream found to have fewer ratings (1–1.4 mg/m^3) as compared to other two streams. It was then concluded that sand mining has a high impact on air quality as it was proven by the dust level found within the mining sections (see Table 1).

3.6 *Riparian Zone*

Although there was no test or measurements conducted for the riparian zone, observations were made and it was noted that large quantity of vegetation was removed during development of sand mining and also along the exploitation of the sand deposit (see Fig. 6). Heavy machinery was also found to destruct vegetation growths within the mining areas and along the travelling, haulage roads.

3.7 *Sieve Analysis*

The grain size of soil is the powerful tool for describing a site's geomorphic setting and determines local sediments transport mechanism. Sieve analysis was conducted

Fig. 6 Removal of vegetation due to sand mining activities

Table 2 Results of sieve analysis for soil classification

Sieve size (Um)	Mass retained (g)	Percentage of mass retained	Cumulative percentage	Per cent passing
2000	0	0	0	100.00
1000	70.38	6.27	6.27	93.73
500	73.63	6.56	12.83	87.17
250	127.83	11.39	24.22	75.78
125	357.5	31.85	56.07	43.93
75	168.44	15.01	71.08	28.92
63	34.37	3.06	74.14	25.86
32	290.18	25.86	100.00	0.0

to identify whether the soil type of the study area was prone to erosion or erosion was only influenced by sand mining. The results for soil classification of the study area are shown in Table 2. It was found that the soil of the study area was considered to be poorly graded as a result of containing particles of nearly same size; however, poorly graded soil is more susceptible to soil liquefaction than well-graded soil. It was then concluded that riverbank collapse and riverbank erosion might be influenced by the type of soil within the study area.

3.8 Geographical Information System (GIS) and Remote Sensing Results and Discussion

In order to validate the results of the study, geographical information system and remote sensing techniques were used to simulate Landsat ETM + Images and normalized difference vegetation index (NDVI) at an interval of five years for the period of 15 years. This section details analysis on vegetation cover status, vegetation cover and sand mining relationship in the Nzhelele valley, regression analysis, Nzhelele valley land use change, accuracy assessment of the Nzhelele valley unsupervised classification and Nzhelele valley change detection.

3.9 Nzhelele Valley Vegetation Cover Status

The vegetation indices from the year 2000 to 2015 in the Nzhelele River Valley were generated using Landsat 7 ETM+ 4 NDVI maps. The NDVI ranges were classified according to what they represent on the ground in order to assess the exact vegetation cover and sand mining relationship. A decreasing negative value indicates no vegetation cover, and an increasing positive value indicates an increase in vegetation cover greenness.

Based on results obtained from NDVI in the year 2000, NDVI values ranging from −0.6 to 0.7 were simulated. It was noted that less vegetation (−0.6 to −0.2) was noted at the upstream section, and this might be the result of sand mining. Owing to that the middle stream was found to consist of moderate vegetation (0.2–0.3) with separated portion of scattered vegetation, the scattered vegetation sections were assumed to be influenced by limited sand mining. Lastly, the downstream section was noted to have heavy vegetation (0.4–0.7); this might be an indication of absence of sand mining within the area.

In the year 2005, the NDVI values were noted to range from −0.4 to 0.8 along the Nzhelele River Valley. However, less vegetation was still noted at the upstream section, and scattered vegetation cover was pounced at the middle section of the river, with high density of vegetation cover in the lower section of the river. It was also noted that sand mining activities were expanding or taking place regularly at the upstream section of the river and gradually occurs of sand mining activities taking place at the middle-stream section of the river in the year 2005 as compared to 2000. It was then concluded that sand mining was highly concentrated at the upstream section during the year 2005, with moderate activities taking place at the middle section of the river. Furthermore, analysis has shown that sand mining activities at the lower section of the river were minimal due to high density of the vegetation cover within the area.

In the year 2010, NDVI values of −0.4 to 0.8 were also simulated; it was noted that much of the sand mining activities were taking place at the upstream section of the river, with less sand mining activities taking place along the middle-stream section of the river, and lastly, heavy vegetation cover was noted at the lower-stream section of the river. Based on the results obtained in 2010, it was then concluded that gradually increase in vegetation cover noted at the middle-stream section between 2005 and 2010 might be due abandon of the section and focusing more on the upstream section.

Lastly in the year 2015, the NDVI values were found to range between −0.4 and 0.8; however, sand mining activities were then noted to extremely taking place along upstream section and middle-stream section of the river with heavy vegetation density at the lower-stream section of the river. This might be an indication that between 2010 and 2015 there was high demand for sand within the areas; due to that, sand mining activities were then conducted along upstream and middle-stream sections of the river as to satisfy the demand. The results are shown in Fig. 7a–d.

3.10 Nzhelele Valley Land Use Change

The effects of sand mining in the valley resulted in the clearance of vegetation cover. However, the presence of vegetation cover resulted in bare land of which sand was being mined over time. During the year 2000, a total of 56% of bare land

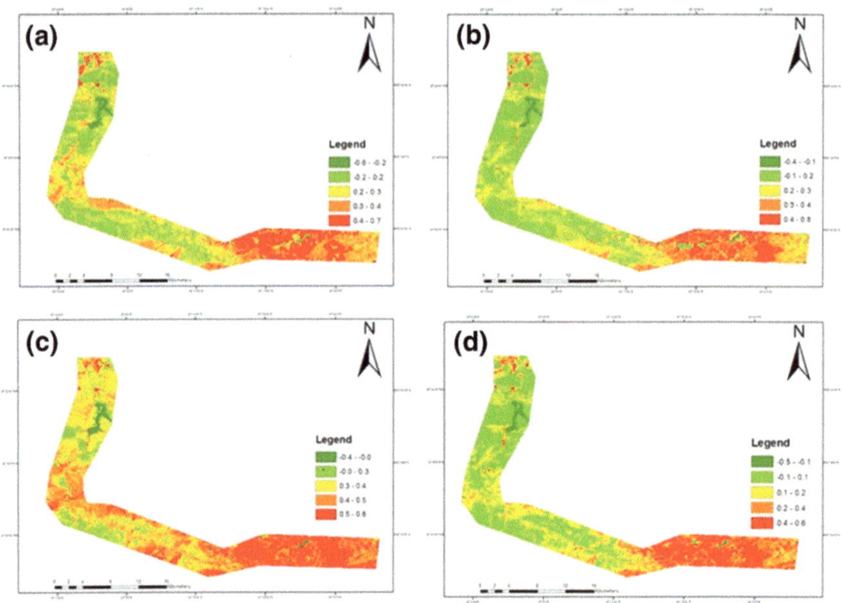

Fig. 7 NDVI classification of Nzhelele valley: **a** year 2000; **b** year 2005; **c** year 2010; **d** year 2015

was observed, this means that sand mining in this area has already been taking place because vegetation was only 41% of coverage in the Nzhelele valley in 2005, and there was a slight increase in vegetation cover by 6% in the valley as vegetation was 47% as compared to the 41% of the year 2000. The bare land decreased to 48% as a result of the increase in vegetation. The effects of sand mining resulted in a competition between the bare land and vegetation in the Nzhelele valley.

A dramatic increase in vegetation cover was observed during the year 2010 as vegetation increased to 86% and bare land decreased to 8%. This is because of politics that arouse in the Nzhelele valley as sand mining was very intense and people were stopped to practise sand mining in the valley. This shows that there might be a gap of years where sand was not mined in the valley thus giving rise to the vegetation cover. However, during the year 2015, the level of sand mining and vegetation cover was slightly the same with vegetation at 49% and bare land at 47%, only a difference of 2%. Figure 8a–d is a graphical presentation of the unsupervised statistics showing the trend of the land cover classes as a result of sand mining in the Nzhelele valley over the period of 15 years. This was to assess the effects of sand mining on land use change in the Nzhelele valley over the period of 15 years.

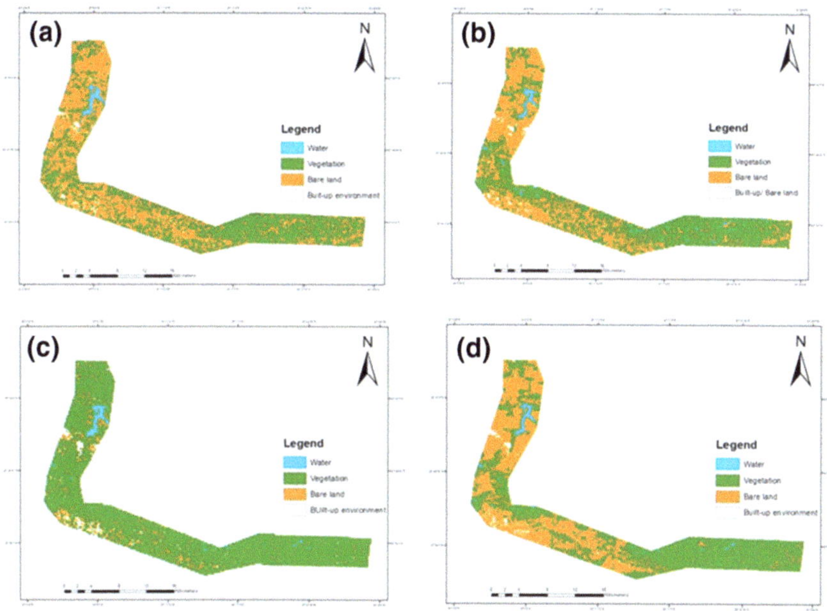

Fig. 8 Unsupervised classification of Nzhelele valley: **a** year 2000; **b** year 2005; **c** year 2010; **d** year 2015

3.11 Accuracy Assessment of the Nzhelele Valley Unsupervised Classification

The study also calculated measures, such as the kappa statistics, overall accuracy, user accuracy and producer accuracy asserted by Garton (2015). Figure 9 shows 160 randomly selected reference points across the study area. The accuracy of the unsupervised classification maps of 2000, 2005 and 2015 was assessed. According to the kappa statistics and the overall accuracy, it is notable that the unsupervised classification of the maps produced realistic results in the study area. The overall accuracy of the classified images is 82% in 2000, 64% in 2005, 81% in 2010 and 84% in 2015. The kappa statistics is 0.79 in 2000, 0.60 in 2005, 0.79 in 201 and 0.82 in 2015. However, the kappa coefficient indicates the percentage of agreement derived after eliminating the percentage of agreement that could be anticipated to take place by a chance. Since kappa coefficient lies on a scale between zero (0), indicating no reduction in error, and one (1), indicating a complete reduction in error, 0.79, 0.60, 0.79 and 0.82 kappa statistics indicate a strong agreement of the unsupervised classification of the study area with minimal errors in the classification.

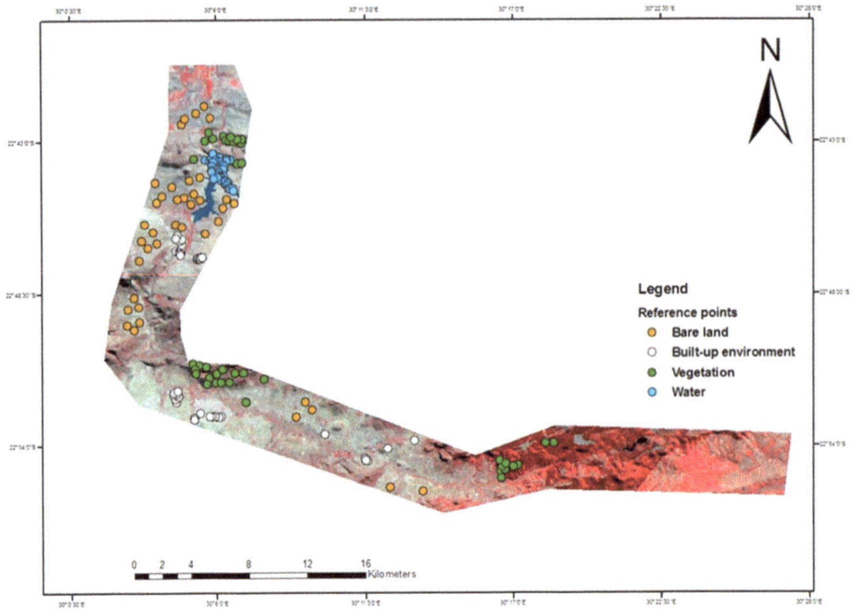

Fig. 9 Randomly selected reference points

3.12 Nzhelele Valley Change Detection

The image differencing in ArcMap was used to produce a change detection map by integrating two Landsats of different years. The Landsats of 2000 and 2015 were combined together to produce the change detection map. According to the difference function, areas that have experienced change between 2000 and 2015 were clearly displayed. The unsupervised image classification together with the raster calculator generated a new image containing the result of only the classes containing the change. The change detection statistics show that the Nzhelele valley has experienced 46% of change as a result of sand mining over the period of 15 years since 2000. Intense change can be observed in the upstream section of the study area where sand mining is very intense. It can be deduced that sand mining has an adverse effect on land use change in the Nzhelele River Valley and sand mining resulted in a competitive relationship between vegetation cover and bare land. Figure 10 shows the change detection map.

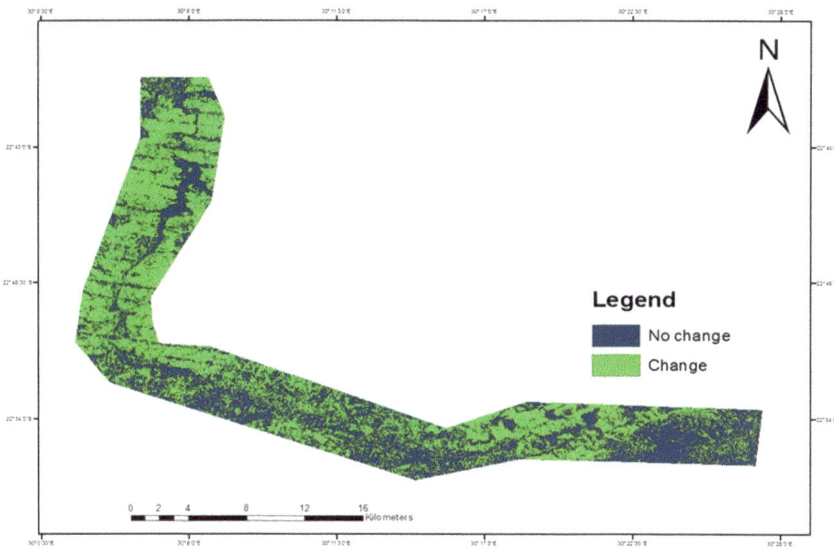

Fig. 10 Nzhelele Valley change detection

4 Conclusions

The study evaluated the environmental impacts of sand mining in the Nzhelele valley using GIS and remote sensing and other statistical methods. The study found that the intensity of sand mining in the Nzhelele valley was 90% in the north-west part of the valley and had adverse effects on the vegetation cover. Sand mining was identified as the only anthropogenic activity having impacts on the vegetation cover. Additionally, spatial variation on vegetation cover was also evaluated as vegetation cover was high and never disturbed by sand mining in the south-east part of the valley. From the findings, the study has shown the significant relationship to be 89% between sand mining and vegetation cover. With such strong relationship, the vegetation cover is expected to decrease if sand mining activities persist in the study area. Results from site observation show that riverbank erosion, depletion in water quality, destruction of the riparian zone, reduction in farms and grazing land and air pollution are the negative impacts associated with sand mining along Letaba River. It can be concluded that these impacts are related to one another because one impact causes the other impact to take place. On the economic point of view, it can be concluded that sand mining at Letaba River is reduction in unemployment rate, crime rate and poverty in South Africa. Sand mining at Letaba River is one of the major creations of income to the villagers, and it also made to simplify the affordability of sand for construction purpose.

References

Brandl, G.: The Geology of the Alldays Area. Explanation: Sheet 2228 (Alldays). Council for Geoscience. 71pp (2002)

Bruvold, W.H., Ongerth, H.J.: Taste Quality of Mineralized Water. J. Am. Water Works Asso. **61**, 120 (1969)

Bumby, A.J.: The geology of the Blouberg Formation, Waterberg and Soutpansberg Groups in the area of Blouberg Mountain. A thesis submitted in partial fulfillment of the requirements of the University of Pretoria for the degree of Doctor of Philosophy. The University of Pretoria (2000)

Carnie, T. Sand mining threatens South Africa coast. [Online] available from www.iol.co.za>business>news>sand. Accessed 21 Nov 2015 (2015)

Chevallier, R. Illegal sand mining in South Africa. SAIIA Policy Briefing. 116 P. 2 (2014)

Chevallier, R. Illegal sand mining in South Africa. [Online] available from www.saiia.org.za>news>new-research. Accessed 9 Nov 2015 (2015)

Dallas, H.F., Day, J.A.: The Effects of Water Quality Variables on the Riverine Ecosystem. A Review. Freshwater Research Unit, Pretoria (1993)

Duffy, L.K., Phyllis, K.W.S.: Effects of total dissolved solids on aquatic organisms: a review of literature and recommendation for salmonid species. Am. J. Environ. Sci. **3**(1), 1 (2007)

Garton, E.O. Simple random sampling. In: Fish and Wildlife Population Ecology, 543 (2015)

Gwimbi, P., Dirwai, C.: Research Methods in Geography and Environmental Studies. Zimbabwe Open University, Harare (2003)

Haupt, C.J., Sami, K.: Letaba catchment reserve determination study. Groundw. Scoping Rep. **2**, 3–15 (2006)

Madise, T. A case study of environmental impacts of sand mining and gravel extraction for urban development. A thesis submitted in accordance with the requirements for the degree of Masters of Science. University of South Africa, Gaborone (2013)

South Africa. Department of Water Affairs and Forestry, South African Water Quality Standards Guidelines. vol. 7, Aquatic Ecosystem (1996)

South Africa. Department of Water Affairs, Luvuvhu/Letaba Water Management Area. Internal Strategic Perspective. Goba Moshloli Keeve Steyn (Pty) Ltd. (2004)

World Health Organization, Guidelines for Drinking Water Quality (2nd ed.). Health Criteria and Other Supporting Information. vol. 2, World health organization, Geneva (1996)

Evaluation of Factors Influencing Slope Instability: Case Study of the R523 Road Between Thathe Vondo and Khalvha Area in South Africa

F. Sengani and T. Zvarivadza

1 Introduction

There are two types of slopes that can be found on the earth's surface, namely natural slopes and artificial or engineered slopes (Abramson and Thomas 2002). Natural slopes are formed usually over a long period of time, through many geological and geomorphological processes such as mountain building, glacial activities, tidal and river activity (Abramson and Thomas 2002). These slopes are stable if the soil has sufficient strength to resist the gravitational forces on the potentially sliding mass. Change in pore water pressure condition, slope geometry or engineering works may cause the natural slope to fail (Wills et al. 2008). Engineered or artificial slopes may be considered in three main categories: embankments, cut-slope, retaining walls and waste tips (Abramson and Thomas 2002).

1.1 Fundamentals of Slope Failure

Debris flow, slumps, mudflows and rock-falls are simply various names given for slope failures. Slope failure is the downslope movement of soil or rock material under the influence of gravity without the direct aid of other media such as water, air or ice (Skinner and Porter 1987). Water and ice, however, are frequently involved in slope failure by reducing the strength of rock and soil and by contributing to plastic and fluid behaviour of soils. Slope failure is a common phenomenon in all high and steep slopes. It can also occur at a very low angle.

F. Sengani (✉) · T. Zvarivadza
School of Mining Engineering, The University of the Witswatersrand,
Johannesburg, South Africa
e-mail: fhatugeorge@gmail.com

E. Widzyk-Capehart et al. (eds.), *Proceedings of the 18th Symposium on Environmental Issues and Waste Management in Energy and Mineral Production*,
https://doi.org/10.1007/978-3-319-99903-6_7

The force of gravity plays a very crucial role in causing slope failure. When the gravitational force acting on a slope exceeds its resisting force, slope failure occurs (Monroe 1994). The resisting forces helping to maintain slope stability include the slope materials strength and cohesion, the amount of internal friction between the grains and any external support of the slope. These factors collectively define a slope's shear strength. Opposing a slope's shear strength is the force of gravity. Gravity operates vertically, but has component acting parallel to the slope, thereby causing instability. The greater the slope's angle, the greater the component of force acting parallel to the slope, and the greater the chance of slope failure to occur (Skinner and Porter 1987). The steepest angle that the slopes can maintain without collapsing is called the angle of repose. At this angle, the shear strength of the slope's material exactly counterbalances the force of gravity. For unconsolidated material, the angle of repose normally ranges from 25° to 40°. Slopes steeper than 40° must be consist of un-weathered solid rock or highly consolidated soil.

1.2 Types of Slope Movement and Instability

There are many types of slope movements, and sometimes they act alone, and sometimes they are combined. The importance of understanding the type of movement is twofold. First, it provides clues to what might be causing the movements, and second, it provides an understanding of the mechanism necessary to design/construct mitigation measures (Skinner and Porter 1987). There are many criteria available for distinguishing types of slope failure: they include velocity and mechanism of movement, character of the rock and soils, and type of slope failures, the shape of the moving mass and water content of the material (Skinner and Porter 1987). Table 1 shows the classification of slope failure.

Table 1 Classification of slope failure (Monroe 1994)

Types of movement		Subdivision/material	Rate of movement
Falls		Rock-falls	Extremely rapid
Slides	Rotational	Slump	Extremely slow to moderate
	Translational	Rockslide	Rapid to very rapid
Flows		Mudflow	Very rapid
		Debris flow	Rapid to very rapid
		Earthflow	Slow to moderate
		Quick clay	Rapid to very rapid
		Solifluction	Slow
		Creep	Extremely slow
Complex movement			Slow to extremely rapid

2 Research Approach

A field investigation was conducted to identify the unstable slopes, types of slope failure, type of material forming the slope and factors that contribute to slope instability of the cut-slope in the study area. Field slope stability assessment, geological mapping and soil sampling were conducted. The laboratory work was conducted towards an understanding of material behaviour, to determine the type of soil forming the cut-slope and how the soil will perform under the imposed conditions such as the presence of water, the rate of rainfall and direction of surface runoff. All these factors affect the behaviour of soils. Analysis of site samples with respect to how conditions would change over time provided an understanding of the existing and future conditions. Understanding of the behaviour of material was imperative for assessment of the stability of the cut-slope.

3 Results

Various factors impacting the road construction along the study area were investigated; a summary of seven case studies is presented in this paper. In addition, the results of the study of soil properties are presented, followed by identification of slope stability and hazardous zones along Road R523.

3.1 Road Construction

Based on the observation and historical information concerning the geographical features of the study areas, it was noted that the road construction in the study areas has resulted in a wide variety of environmental impacts along the road R523. It had altered the surface drainage patterns and increased runoff generation. The improper road construction in the study area causes the concentration of water at the foot of a cut-slope, and this provides additional weight to the slope which in turn increases the driving force which causes slope failure. The alteration of surface drainage patterns results in the increase of speed of surface runoff and therefore a high probability of water washing soil down to the road, blocking the surface drainage ditch. As a result, the foot of the slope became waterlogged. When the soil is over-saturated with water, slope failure occurs.

The observed significant geomorphic impacts of road construction along the road R523 include the destabilization of side-cast material downslope, also making the slope steeper than before. Thus, the road construction has resulted in undercutting of natural slopes that were at the angle of repose, creating new steeper cut-slopes that are unstable and susceptible to failure. When the slope steepness is increased, the shear stresses in the soil within the slope also increases and reduces stability, resulting in slope failure (Abramson and Thomas 2002).

3.2 Case Study of a Slope Failure Area

During field observation, the evidence of a previous slope failure was noticed (see Fig. 1) in the study area where slope failures are not active. There were also mitigation measures implemented after slope failure occurred. In this case, the gabion wall to prevent future slope instability was built. The gabion wall was noted to be effective where two-tiered steps were employed. It was found to be ineffective where less than two-tiered walls were employed. The ineffectiveness of the gabion was noted by through a slope failure that had been reactivated. As a result, the failure of the gabion was noted to have led to the road interruption due to soil slides which blocked the road.

Furthermore, analysis of factors that contributed to the slope failure along the study areas has indicated that water pressure also plays a major role in the weight of the soil. The previous study by Duncan and Wright (2005) has shown that water that fills the cracks developed on the surface of the slope has high possibility to decrease the shear strength of the soil as pore pressure in the slope increases.

Fig. 1 Slope failure at R523 road

Based on the observation, in most of the areas where slope failure was more pronounced, several cracks were noted and a number of short streams were also noted. It was believed that the slope failure in the study areas was highly influenced by the presence of a large quantity of water flowing around the areas. It was also observed that most of the slope failure took place during the rainy season.

3.3 Soil Physical Properties

3.3.1 Particle Size and Size Distribution

Based on the laboratory analysis, the soil particle size distribution of the study areas was found to be significantly richer in silt and clay contents (see Table 1). The clay content of the soil varied between 27.89 and 40.95%, silt content between 40.03 and 51.96%, whereas sand content ranged from 13.94 to 26.97%. Furthermore, analysis has shown that the soil texture of the study area can be classified as a silty clay (S1), silty clay loam (S2) and clay loam (S3, S4, S5 and S6). The soil was classified using the US Department of Agriculture System (USDA) classification system. Due to the above-mentioned characteristic of the soil in the study area, it was then concluded that the area is therefore prone to land-sliding. This is because when clay soil is over-saturated due to heavy rain, the soil is usually subjected to swelling and this ultimately results in land-sliding phenomenon. Furthermore, when these soils get dried, it is mostly subjected to shrinkage which will result in a change in the volume of the soil mass; thus, the soil fragments of varied dimensions will be subjected to movement (Table 2).

3.3.2 Atterberg Limits

The Atterberg limits are a set of index tests performed on fine-grained soils to determine the relative activity of the soils and their relationship to moisture content

Table 2 Physical properties of soil within the study

Site no.	Sample no.	Particle size distribution			Atterberg limits		
		Sand % (>63 μm)	Silt % (63–2 μm)	Clay % (<2 μm)	LL	PL	PI
1	Sp.1	13.94	44.94	40.95	39.74	30.76	8.98
2	Sp.2	15.92	51.96	33.86	38.32	29.44	8.88
3	Sp.3	25.74	45.98	27.89	52.39	42.42	9.97
4	Sp.4	24.02	43.89	32.01	56.36	37.82	18.54
5	Sp.5	30.01	40.03	29.98	54.15	43.19	9.58
6	Sp.6	26.97	41.87	30.93	53.39	37.58	15.81

(Wills et al. 2008). The laboratory results of measured liquid limit (LL), plastic limit (PL) and plastic index (PI) of studied soil samples were variable. The LL found ranged from 38.2 to 56.38%, PL from 29.44 to 44.56% and PI ranged from 8.88 to 12.6. The plastic limit is the division between a semi-solid and plastic state for the soil material. Therefore, the plastic limit of 44.56% was found to be the highest water content at which the soil begins to behave in a plastic manner. The liquid limit is the division between plastic and liquid state for the soil material. Therefore, a liquid limit of 38.2% was found to be the lowest moisture content at which the soil will flow at very low shear forces.

Based on the results, it was noted that if the water content of the soil in the study area was reduced below the plastic limit of 44.56%, the soil behaved as a semi-solid material, thus increasing the resisting force on the soil, and therefore soil failure was not going to occur. If the water content of soil slope's surface increases above the liquid limit, which is likely to occur in the summer season, in this case, the soil turns to behave as a viscous fluid. As a result, such soil has a high possibility of experiencing failure.

3.4 Slope Stability and Hazardous Zones Along Road R523

One of the guiding principles of slope stability assessment stated that 'the past is the key to the future'. When evaluating landslide hazards future slope failures could occur as a result of the same geologic or geotechnical, geomorphic and hydrologic situations that led to past and present failures (Hadjin 2002).

Based on this assumption, a quantitative classification system was used to determine and identify stable and hazard zones in the study area. This system was based on the consideration of five main factors that are believed to substantially affect the stability of slopes. The slope stability assessment was performed on seven

Table 3 Results of slope stability assessment (input data)

Site no.	Parameters				
	Geologic formation	Slope aspect	Precipitation (Avg. Ann) (mm)	Land-use	Height (m)
1	Silty clay	South-east	120.64	Bush	6.70
2	Silty clay loam	North-east	120.64	Less-dense bushland	8.20
3	Clay loam	South	120.64	Woodland	27.90
4	Clay loam	North	120.64	Woodland	34.5
5	Clay loam	South	120.64	Woodland	17
6	Clay loam	South-east	120.64	Woodland	15.60
7	Sandstone	North-east	120.64	Less-dense bushland	36.50

Table 4 Results of total rate state of stability

Site no.	Position		Factors					
	Latitude	Longitude	Geologic formation	Slope aspect	Precipitation (Avg. Ann) (mm)	Land-use	Height (m)	Total score
1	22° 56′ 48″	30° 21′ 19.8″	5	15	10	15	16	61
2	22° 55′ 59.6″	30° 20′ 34.2″	5	15	10	15	16	61
3	22° 55′ 58.2″	30° 20′ 27.7″	5	10	10	10	4	39
4	22° 55′ 56.1″	30° 20′ 23.7″	5	8	10	10	4	37
5	22° 55′ 41.4″	30° 20′ 9.5″	5	10	10	10	13	48
6	22° 55′ 42.7″	30° 19′ 38.9″	5	15	10	10	13	53
7	22° 55′ 37.1″	30° 17′ 58.3″	15	15	10	15	4	59

cut-slopes along the road R523, and the results are summarized in Tables 3 and 4. The overall slope stability was obtained by summing the rates of all contributing parameters, these results are shown in Table 4, and five classes of stability and hazard can be found in Table 3.

Based on the qualitative classification system results, slopes at site no. 3, 4, 5 and 6 were classified as unstable slopes which made the most hazardous section along the road R523 and these slopes need to be stabilized to improve their stability. It was noticed that the slopes at these sections had a high probability of experiencing failure. This was evident due to that almost all slope failure in the study area was occurring in the area of previous slope failure.

The slopes at site no. 1, 2 and 7 were classified as fairly stable slope after summing all five factors believed to affect their stability. These slopes also fall within the class of slope with moderate hazard, based on Table 3. The stability of these slopes needed some improvements as to prevent the possibility of future slope failure. The slope at site no. 7 was totally different from other slope due to being entirely composed of fractured sandstone while others slopes are composed of soil.

4 Conclusions

It was concluded that based on the recorded data and field observation, the slope failure was prone to occur during heavy rainfall. It was noted that the rainfall turns to trigger slope failure due to the increase in water pressure within the soil pores. Furthermore, analysis has shown that the slope failures identified in the field were rotational slump failures, and they occurred close to the road. It was then concluded that the natural slopes were disturbed during road construction since the natural drainage pattern was altered. This contributed to slope failure. During this study, four factors, namely degrading of slope material, water pressure in the cracks on the surface of the slope, road construction and heavy rainfall, were identified as main factors that affected the stability of cut-slope in the study area. Moreover, the laboratory studies have shown that the study areas consist of most clay soil, which is commonly known for the high possibility of sliding and shrinkage characteristic. It was then concluded that the soil properties of the study area also played a major role in the failure of the slopes.

References

Abramson, L.W., Thomas, S.L.: Slope Stability and Stabilization Methods, 2nd edn, pp. 1–48. Wily, New York (2002)

Duncan J.M., Wright, S.G.: Soil Strength and Slope Stability. Wiley, New Jersey (2005)

Hadjin, D.J.: New York State Department of Transportation Rock Slope Rating Procedure and Rock-fall Assessment, Transportation Research Record, 1786, Paper number 02-3978 (2002)

Monroe, J.S.: The Changing Earth: Exploring Geology and Evolution, pp. 269–289. West Publishing Company, New York (1994)
Skinner, B.J., Porter, S.C.: Physical Geology, pp. 205–237. Wiley, New York (1987)
Wills, C.J., Manson, K.D., Brown, C.W., Daveport and Domrose, C.J.: Special report, Landslides in the highway 1 corridor: geology and slope stability along the big sur coast between point lobos and san carposoro creek, montery and san luis Obispo counties, California, Department of conservation, California geological survey (2008)

Environmental Issues with Best Management Practices in Energy and Mineral Production

Gurdeep Singh

1 Introduction

Minerals play an important role in the economic development of the country as minerals are the basic raw materials to promote the growth. The development and extent of judicious utilization of mineral resources add to the index of growth of a nation and its people. The mineral industry in India is reckoned not only as an important contributor to the country's GDP and foreign trade, but also as one of the major industries that absorb a considerable amount of the country's working population. The Indian mining industry provides employment to about 1.1 million persons. India's mining industry is projected to touch over \$50 billion (about Rs. 282,966 crores) accounting for about 2.5% of the GDP in the next 4 years, the latest report said (Ministry of Mines 2016; Saxena et al. 2005). This industry is spread almost all over the Indian Territory and has operations in some of the remotest areas of the country, where it can claim itself to be the sole leader of infrastructure development. The mining leases occupying about 0.7 million hectares which is 0.21% of the total land mass of the country. This industry operates more than 2600 mines which consist of 574 coal mines, 2054 metalliferous mines and a score of other small mines (Ministry of Mines 2016; GOI's Hydrocarbons Vision 2025 Report; BP Statistical Review of World Energy 2005). India produces 87 minerals out of which 4 are fuel minerals, 10 metallic, 47 non-metallic and 23 minor minerals. The Indian economy to a great extent depends on the value of the minerals produced, as these represent a major portion of the raw materials for the nation's industrial activities. The report on metals and mining pointed out that India has immense natural resources and is ranked among top 10 globally for production, second in barytes, chromite and talc/steatite/pyrophyllite, third in coal and lignite,

G. Singh (✉)
Department of Environmental Science and Engineering, Centre of Mining Environment, Indian Institute of Technology (Indian School of Mines), Dhanbad, India
e-mail: gurdeep@iitism.ac.in

E. Widzyk-Capehart et al. (eds.), *Proceedings of the 18th Symposium on Environmental Issues and Waste Management in Energy and Mineral Production*,
https://doi.org/10.1007/978-3-319-99903-6_8

fourth in iron ore and kyanite/sillimanite, fifth in manganese ore, steel (crude) and zinc, sixth in bauxite, eighth in aluminium and tenth in magnesite of the world's resources (Saxena et al. 2005).

Indiscriminate and unplanned mining causes irreversible damage and deterioration of natural resources. Mining activities affect surrounding, i.e. air, water, soil, land, biological diversity. The environmental impacts of mining activities may have short-term and long-term implications. Guidelines for taking necessary precautions before, during and after mining operations are laid down to ensure sustainable development. The role of mining in sustainable development is one issue that decision makers and resource managers have wrestled with for decades. Mining is one of those activities that really connect issues relating to people, development and the environment. The negative impact of mining on health, land, water, air, plants and animals and other aspects of society can be reduced by careful planning and implementation of mining activities. It is essential to strike a balance between mineral developments on the one hand and the restoration of the environment on the other hand.

1.1 Environmental Issues

The mining operations like drilling, blasting, extraction, transportation, crushing and other associated activities are carried out in underground and opencast mines (Singh 2005a; Saxena and Ghosh 1996). Mining operations damage the environment and ecology to an unacceptable degree, unless carefully planned and controlled. There is a need for balance between mining and environmental requirements.

The various impacts of mining on environment and their mitigation measures are presented in this chapter.

1.1.1 Impact of Mining on Air Quality

Air pollution in mines is mainly due to the fugitive emissions of particulate matter and gases including methane, sulphur dioxide, oxides of nitrogen and carbon monoxide. Most of the mining operations produce dust. The major operations producing dust are drilling, blasting, hauling, loading, transporting and crushing (Singh and Puri 2004). Basically, dust sources in mines can be categorized as primary sources that generate the dust and secondary sources, which disperse the dust and carry it from place to place called as fugitive dust.

Opencast mining is more severe air pollution problem in comparison with underground mining. High levels of suspended particulate matter increase respiratory diseases such as chronic bronchitis and asthma cases, while gaseous emissions contribute towards global warming besides causing health hazards to the exposed population. The uncontrolled dust not only creates serious health hazard but also affects the productivity through poor visibility, breakdown of equipment, increased maintenance cost and ultimately deteriorates the ambient air quality in and around the mining site. The dust can also pollute nearby surface waters and stunt crop growth by shading and clogging the pores of the plants. Besides polluting the environment, the generation of dust means the loss of fines, which act as road surface binders.

Problem with Greenhouse Gases, Acid Rain and Ground-Level Ozone
The key environmental challenges facing the coal industry are related to both coal mining and the use of coal—greenhouse gases, acid rain and ground-level ozone, issues which can be local, regional and global in their impacts.

The greenhouse effect is a natural phenomenon, which refers to the increase in the earth's surface temperature due to the presence of certain gases in the atmosphere. There is concern that this natural phenomenon is being altered by a greater build-up of gases caused by human activity. This is known as the enhanced greenhouse effect. The combustion of coal, like that of other fossil fuels, produces CO_2, a gas that is linked to global warming through the greenhouse effect.

The combustion of coal produces gaseous emissions of sulphur dioxide (SO_2) and nitrous oxides (NO_x) that are responsible for the production of 'acid rain' and 'ground-level ozone' (Singh and Puri 2004; Singh 2005b). Acid rain occurs when SO_2 and NO_x gases react in the atmosphere with water, oxygen and other chemicals to form acidic compounds. Ground-level ozone (O_3) is mainly responsible for smog that forms a brown haze over cities. Ground-level ozone is formed when NO_x gases react with other chemicals in the atmosphere and is enhanced by strong sunlight. Emissions of SO_2 and NO_x are termed trans-boundary air pollution because the environmental impacts from the production of these gases are not restricted by geographical boundaries.

1.1.2 Impact of Mining on Water Regime

Disturbance to Hydrologic Regime
Mining and its associated activities not only use a lot of water but also affect the hydrological regime of the district and often affect the water quality. Large and deep opencast mines usually have great impact on the hydrologic regime of the region. The major hydrological impact of a large and deep opencast mine, however, is on the groundwater regime of the region. The water seeping into the mine and collected in the mine sump is partly used up in the mine, and the excess amount is discharged into the surface drainage system. The water used up in the mine for spraying on haul roads, conveyors, at loading and unloading points, bunkers, etc., is lost by evaporation. A deep mine is likely to have longer haul roads requiring more spraying water. The water used for green belts and plantation areas is also lost by evapotranspiration. Many areas of the country are faced with the problem of

over-exploitation of groundwater resources resulting in alarming lowering of water table (Dutta et al. 2002; Ghosh and Singh 2003). Therefore, a lot of care has to be taken in estimating the water need and the mines of future are likely to be subjected to a lot of constraints on water use and discharge.

Acid Mine Drainage
Acidic water results in severe water pollution problems. Acid mine drainage (AMD) refers to distinctive types of waste bodies that originate from the weathering and leaching of sulphide minerals present in coal and associated strata. Environmental effects of AMD include contamination of drinking water and disrupted growth and reproduction of aquatic plants and animals. Effects of AMD related to water pollution include the killing of fish and loss of aquatic life and corrosion of mining equipment and structures, such as barges, bridges and concrete materials.

AMD is the most persistent pollution problem in mines of North Eastern Coalfield. Generally, water quality characteristics of acidic mine water reflect high acidity and high hardness along with high iron and sulphate contents (Singh 1988, 1999). Various toxic trace/heavy metals become soluble in acidic water and may be presenting significance to concentration levels depending upon their availability in the source material. Fortunately, the considerable majorities of coal mining areas are safe and only in a few localized areas problem of AMD exists (Banejee and Singh 1993). AMD cripples the economy of mines due to compliance of stringent environmental standards and involves huge cost burden in its management.

1.1.3 Impact of Mining on Land

Irrespective of the type of mining used for extracting coal, mining invariably results in enormous land disturbance—e.g. large-scale excavation, removal of topsoil, dumping of solid wastes, cutting of roads, creation of derelict land. The mining industry, in general, is reluctant to rehandle overburden material for economic reasons, but in a few cases it has been planned to rehandle the material to fill the voids created at the end of mining, and it is expected that the practice will become more widespread in future.

Opencast mining has more potential impact on land than underground mining. With improved technology, opencast coal mining is being used extensively because of its cost-effectiveness and productivity though it results in large-scale land disturbance. Although underground mining has considerably less impact than opencast mining on land, it causes enough damage through subsidence as observed in Jharia and Raniganj coalfields (Saxena and Ghosh 1996; Bansal 2004). The surface subsidence inflicts severe damages to engineering structures such as highways, buildings, bridges and drainage besides interfering with groundwater regime.

1.1.4 Impact of Noise and Vibrations from Mining

A cumulative effect of all mining activities produces enormous noise and vibrations in the mining area, which constitutes a source of disturbance. The availability of large diameter, high-capacity pneumatic drills, blasting of hundreds of tonnes of explosive, etc., is identified as noise-prone activities. Input crushing system with mobile crusher and large capacity materials handling plants are being installed to facilitate speedy handling of large quantities. All these activities are major sources of noise and vibrations in and around the mining complexes (Saxena et al. 2005).

The obvious implication of noise is, of course, the potential for noise-induced hearing loss. In addition, noise produces other health effects, influences work performance and makes communications more difficult. Besides, the fauna in the forests and other areas surrounding the mines/industrial complexes is also effected by noise and it has generally been believed that wildlife is more sensitive to noise and vibrations than the human beings.

1.1.5 Impact of Coal Mine Fires

A number of coal mines in the country are affected by fires leading to the steady destruction of precious energy resource. The reason for mine fires presumably involves the phenomenon of spontaneous heating through two interrelated processes, viz. the oxygen coal interaction or oxidative process and the thermal process. If it remains uncontrolled, the fire could spread further through interconnected pathways and fissures in the strata. It is estimated that about 10% of total national coal resources are in the fire-affected areas (Saxena et al. 2005; Singh 2005b).

Mine fires give rise to several environmental problems besides safety hazards and economic losses. Apart from direct losses due to burning of coal, the other associated hazards encountered are: (i) gas poisoning, (ii) difficult geo-mining conditions, (iii) sterilization of coal, (iv) hindrance to production, (v) explosions, (vi) damage to structure and adjacent properties, etc.

1.2 Social Issues

Coal mining, despite the very substantial benefits they bestow on society, stirs strong emotions. A great ongoing social challenge for the mining industry is sustainable development and community acceptance of its role in society. The problem of mining-induced displacement and resettlement (MIDR) poses major risks to societal sustainability (Bhattacharya 2003; Saxena and Pal 2000).

(i) *Landlessness*: MIDR raises the significant risk of landlessness by removing the foundations upon which productive systems, commercial activities and livelihoods are articulated.

(ii) *Joblessness*: The ethnic people living in the designated areas depend generally for their livelihood on the land. Since in mining areas the land is taken for mining and associated activities, these people lose their livelihood.
(iii) *Homelessness*: Defined as the 'loss of house plots, dwellings and shelter'. For many people, homelessness may be only temporary, but in poorly executed displacements, it remains chronic.
(iv) *Risk of marginalization*: The risk of marginalization threatens displaced individuals and entire communities as they slip into lower socio-economic status relative to their local areas.
(v) *Changes in population dynamics*: All the manpower required for mining and associated activities comes from outside as such trained manpower is usually not available in an ethnic population.
(vi) *Cost of living*: Increased industrial and economic activities generate more money and increase the buying power of the people directly and indirectly associated with these activities. This leads to an increase in the cost of living.
(vii) *Health risks*: The already marginal health status of displaces is worsened by the stress and trauma of moving. Recurring problems are reported with resettled populations gaining access to safe potable water and safe sanitation; increased diarrhoea, dysentery and epidemic infections often result.
(viii) *Disruption of formal educational activities*: Risk occurs in the disruption of education and routine socialization. Displacement and relocation often cause a significant interruption in the functioning of schools and in child access to education during the year of transfer or for longer periods of time.
(ix) *Addictions*: Increased economic activities and affluence bring in more addictions in the society. In the tribal areas, the ethnic people may also get affected by additional addictions.

Mining companies must accept the right of landowners to negotiate access to their land, to determine whether or not exploration or mining takes place there. If access is given, they must be able to negotiate conditions, such as the preservation of sacred sites, access to traditional hunting grounds, proper resettlement and rehabilitation of those who have to be moved and the determination of compensation packages.

Account should be taken of alternative plans proposed by the affected people, who must be allowed to identify suitable resettlement sites. Where those being displaced have agriculture as their primary source of income and livelihood, every effort must be made to replace land with land (Ministry of Environment and Forests, Government of India 2003). If suitable land is not available, other strategies built around opportunities for employment or self-employment should be used. Relocated people must receive legal land titles for their resettlement plots, whether these are house plots or agricultural land.

1.3 *Example of Best Management Practices*

Societal Development in Coal Sector (Coal India Limited)
Mining-induced displacement and resettlement (MIDR) increased substantially since the 1970s as the country's coal production shifted from underground to opencast mining. The issue has gone beyond economics and environment; local NGOs, such as Operations Research Group (ORG), a consultant of Coal India Limited (CIL), reported that MIDR is creating a pattern of 'gross violation of human rights' and 'enormous trauma in the country (Krishnamurthy 2004)'. By the mid-1990s, Resettlement and Rehabilitation (R&R) Policy of CIL has been designed to ensure that affected people improve or at least regain their former standard of living and earning capacity after a reasonable transition period (Bhattacharya 2003; Singh and Gupta 2005). Coal India Limited implemented the Environmental and Social Mitigation Project (ESMP) in 25 selected opencast mines with World Bank funding during 1996–2002. Environmental and Social Mitigation Project (ESMP) aimed to mitigate adverse effect of coal mining on environment and people affected by such activities. ESMP consisted of two components (Saxena et al. 2005; Saxena and Pal 2000):

- Environmental component—implemented through Environmental Action Plan (EPA).
- Social component—implemented through Rehabilitation Action Plan (RAP) and Indigenous Peoples' Development Plan (IPDP).

Environmental Action Plan (EAP) includes domestic effluent treatment plant, workshop effluent treatment plant, mine water discharge sedimentation plant, dust suppression majors, tree plantation, OB dump reclamation, topsoil storage and spreading for bio-reclamation, environmental monitoring.

Rehabilitation Action Plan (RAP) includes shifting of villagers affected by mining, resettlement and rehabilitation of project-affected families (PAFs) by giving a plot of land in well-developed resettlement sites or a lump sum package to settle at a place at their choice. The PAPs are also trained in different trades for their economic rehabilitation.

Under Indigenous People Development Plan (IPDP) villages falling within 1 km area from the leasehold of the mines are considered. Activities under IPDP include (1) development of community infrastructure like school building, community hall, dispensary building, village roads, school furniture, wells, and tube wells, (2) community activities like Mahila Mandal, youth club, self-help groups, sports, cultural programmes, (3) training and capacity building, training for self-employment, non-formal education, etc.

Overall environmental management improvement has been taking place with the implementation of state-of-the-art environmental management schemes particularly under Environmental and Social Mitigation Project (ESMP) of CIL.

2 Augmentation of Pumped Out Mine Water for Sustainable Development of Coal Mining Regions

The need to provide potable water supply in the densely and thickly populated highly industrialized, water-scarce coal mining areas of India cannot be over-emphasized. Mining populations in these regions have been facing acute shortage of water supply, and in some places, people have to walk a few kilometres to collect the water for their daily needs. Epidemics also occur due to water-borne diseases, due to the non-availability of a properly treated water supply.

Millions of gallons or mine water are being discharged daily from underground coal mines. This valuable water resource of generally acceptable quality becomes contaminated with various domestic and industrial trade effluents and subsequently is just wasted while putting on an enormous energy cost burden on the underground mines (Banejee and Singh 1993; Singh and Gupta 2005). This study highlights that this underground mine water can be augmented to meet various water supplies, particularly for drinking purposes. Treatment of this water resource at some places is also presented (Singh 2005b). A simple scheme for augmentation of underground mine water for potable purpose is also suggested.

2.1 Water Demand in Coal Mining Areas

There has been an acute shortage of water in many places within the coal mining areas of India, and one such example is Jharia and Raniganj coalfields in eastern India. The quantity of water available for an individual for day-to-day use for various purposes along with the corresponding WHO recommended quantity is presented in Table 1.

2.2 Water Availability Status in Jharia Coalfield

The population of Dhanbad town and the Jharia coalfield is at present at the level of about 15.0 lakh. The potable water demand for an estimated population of 15.0 lakh at 30 GPCD (135 LPCD) works out to 45 Mgd. Considering 10% for loss in distribution and 25% water demand for private industries, the total demand comes to 49.5 Mgd. The user-wide break up of demand is as follows:

Table 1 Quantity of water available for an individual for day-to-day use

S. No.	Purpose	WHO's recommendation litre/person/day (A)	Quantity available litre/person/day (B)	Per cent shortage (%) (C)
1	Drinking/cooking	15	8.5	43.3
2	Bathing/personal washing	60	12.0	80.0
3	Utensils washing	15	5.0	66.6
4	Cloth washing, washing	20	7.5	62.5
5	House washing	10	5.0	50.0
6	Flushing/refuse disposal washing	60	15.0	40.0
7	Garden	10	–	100.0
8	Wastage	20	7.0	65.0
Total		210 L.	60.0 L	71.42

$C = \frac{A-B}{A} \times 100\%$

(a) BCCL employees—19.5 Mgd
(b) Rehabilitation colonies (non-BCCL)—23.0 Mgd
(c) Dhanbad and Baghmara towns—7.0 Mgd.

The Mineral Area Development Authority (MADA) which is the 'sole source' of water supply to the non-BCCL population is supplying water to Dhanbad and Baghmara towns also. The industrial water demand of the Jharia coalfield has been assessed as 34 Mgd.

On the completion of the Jharia Reconstruction Scheme, the underground mines shall be pumping 61 Mgd. The domestic water demand comes to 49.5 Mgd, and the industrial water demand has been assessed as 34 Mgd. The total demand comes to 83.5 Mgd, and the availability would be only 61 Mgd; thus, the whole of the underground pumped out mine water can be treated for potable purposes as this is of good quality except hardness. This will not only meet potable water demand considerably but also cut down the treatment costs under existing conditions. The rest of the water demand will be met from the River Damodar and its tributaries. Even though the underground mine water is fully taped, it does not meet the total demand and cost has to be incurred for abstraction of water from surface water bodies to meet the rest of the demand. As the River Damodar and its tributaries are rain-fed, the flow becomes very negligible in the summer. The water supply becomes a great problem during the summer. Thus, in such a condition mine water is the ideal source which can be augmented to meet the various water supplies particularly for domestic purposes which is in accordance with the National Water Policy. The National Water Policy emphasizes the need to augment the water availability from such sources and to meet various demands of the local people.

2.3 Mine Water Quality Classification

A classification of underground mine waters, as arrived at on the basis of analysis of mine waters from various coal mines of eastern India, is given in Table 1. General water quality characteristics under each category are summed up in Table 1. Most of the mine waters fall under category A2 which have high to medium concentrations of dissolved solids, hardness, sulphate with low to medium concentration of suspended solids. Biochemical parameters (BOD, COD) and toxic metals are present at insignificant levels. Waters of category A1, which are slightly alkaline in nature, differ from those of category A2 in that they are not hard and contain comparatively low concentration of dissolved solids and sulphate. Chloride concentrations of both the categories are not very different and are generally at an acceptable value. Mine waters of category A3 contain very high concentration of dissolved solids particularly sulphate and chloride with very high hardness content, and their occurrence is rare and limited to some of the worst affected mine fire areas. All the mine waters revealed bacteriological contamination, and special attention needs to be paid to disinfection.

2.4 Classification of Mine Waters Based on Hardness

Water samples collected from the underground mines were hard to very hard which is revealed from Table 2. About 75% of the samples were found to be lying in the very hard category. This is the main cause of the underground mine waters non-acceptance among the public, particularly due to difficulty in cooking and consumption of lots of soap for washing and bathing purposes. The release of phosphates from the soaps and detergents consumption is also contributing towards the eutrophication of confined water bodies. Further, hardness content in the mine waters also results in scale formation.

2.5 Pumped Out Mine Water Treatment

Mine water in general is treated mainly by conventional methods using aeration followed by flocculation with rapid and slow mixing and sedimentation. The water enters the rapid gravity filters after which it is disinfected by chlorination and supplied to the consumers. In the case of mine water with significant hardness (>300 mg/l), softening by lime soda process or ion-exchange process is additionally practiced. Addition of coagulant in excess to the mine water will hasten flocculation and increase bactericidal efficiency; thus, it is economical to use the minimum quality of coagulant and to depend on chlorination for bacterial safety of the water.

Table 2 Classification of mine waters

Parameters		Category of the mine waters		
		A1	A2	A3
Water quality parameters				
A	pH	7.5–8.5	7.0–8.5	6.0–8.0
B	Alkalinity (m/l $CaCO_3$)	200–7000	100–500	10–30
C	Hardness (mg/l $CaCO_3$)	50–150	200–1500	1000–2000
D	Total dissolved solids (mg/l)	300–700	500–1500	1500–2500
E	Total suspended solids (mg/l)	5–30	12–50	10–50
Biochemical parameters				
A	DO	5.5–8.0	5.0–8.0	5.0–8.0
B	BOD	<1	<1–2	<1–2
C	COD	8-30	15–50	15–60
Element or ion content				
A	Calcium (mg/l)	10–50	50–200	100–150
B	Magnesium (mg/l)	10–30	40–180	80–100
C	Total iron (mg/l)	0–2	0–5	0–5
D	Sulphate (mg/l)	10–100	100–800	1000 and above
E	Chloride (mg/l)	18–40	2245	80–400
F	Bacteriological contamination coliform organism (MPN/100 ml)	0–180	6–180	0–18

In the mining areas, there are a number of collieries—the workings of which facilitate the pumping of millions of gallons of water which is being discharged putting on enormous costs on energy for pumping. Some of the collieries have their own treatment plants in the vicinity of the mine for the treatment of mine water and supplying it to the mining community. The schematic layout of such treatment plants is given in Fig. 1. Many of these units have not been observed in proper operating conditions. In many of the cases, the filler beds were found to be choked and the quality of water obtained after treatment was not of the required standard. In some cases, the treated water had either excess or deficient of free residual chlorine which is objectionable for the aesthetic and public health point of view.

3 Developing Potential Land-Use/Land-Cover Area in the Talcher–Angul Region for Suitable Locations of Future Industries Through Remote Sensing and CIS

The Talcher–Angul region of Orissa is a fast developing area. The area is located in the Angul, Dhenkanal, Orissa. Geographically, it is bounded by latitudes 20° 30′ to 21° 15′ north and longitudes 84° 45′ to 85° 35′ east, encompassing a total area of

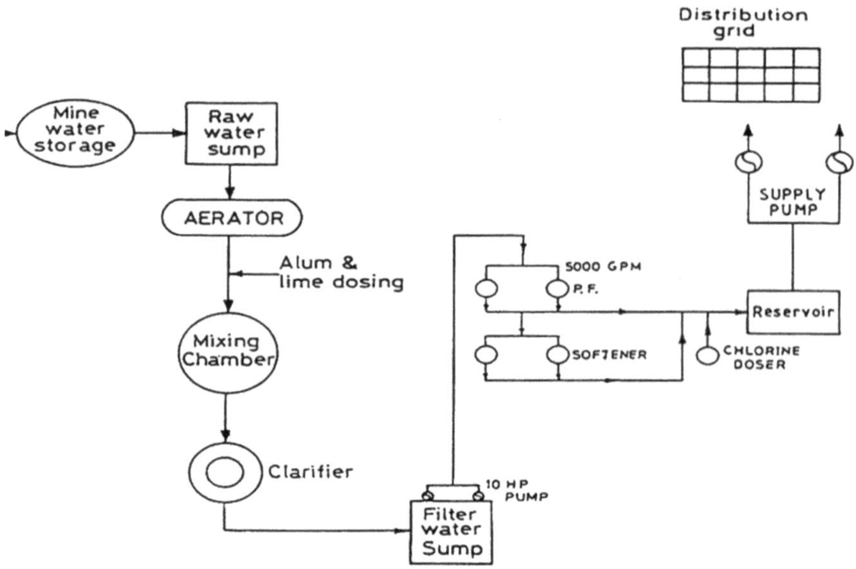

Fig. 1 Schematic layout of water treatment plant in practice at some of the coal mining regions

about 5028 km^2. The Talcher coalfield is one of the oldest coalfields in the country and has been contributing significantly to the country's need in power generation (Singh et al. 2002). This is emerging as a major energy centre, and there is a proposal of setting up 14 new power plants in the region in view of the potential of easy availability of coal coupled with water resources. The Government of Orissa is much interested in setting of new power plants in this region. This gigantic venture requires land and other amenities such as water, transport and other facilities. A good amount of land has already been degraded due to mining, deforestation and urbanization. Requirement for barren lands for setting up these power plants keeping all environmental, economic and sociological needs in mind need not be over-emphasized (Singh et al. 2009). The objective of the study revolves around setting up the future plans for this entire region which includes a proper land-use classification and identification of accessible barren lands throughout the region. As a part of the study, extensive use of remotely sensed data has geared up the task and fetched very good results in strategic planning and management of the region.

3.1 Data Considered: IRS-P6/1C Satellite Imagery (2007)

False colour composite (FCC) satellite imagery of LISS-III and its corresponding digital satellite data were used to prepare recent land-use map of the study area. The

LISS-III data in four bands, i.e. Band 2—green, Band 3—red, Band 4—near infrared and Band 5—mid infrared, acquired by the IRS-P6 satellite on 13 December 2007 (Path no. 105 and Row no. 57, 58) and 18 December 2007 (Path no. 106 and Row no. 57, 58), were used. FCC hard copy satellite imagery on 1:50,000 scales in 12 coloured photographic sheets were prepared by National Remote Sensing Agency (NRSA), Hyderabad. These were geo-coded, i.e. rectified to match the Survey of India topographic maps, free of cloud cover with 12.5 m spatial resolution. The satellite imagery mosaic of the study area (merged satellite data of all the 12 satellite imageries covering the study area) is given in Fig. 2.

3.2 Topographic Maps

Survey of India (SOI) topographic maps on 1:50,000 scale (surveyed between 1973 and 1981) were used as the 'base map' for preparing the land-use/land-cover map of the study area. The study area is covered by 12 SOI topographic maps either in full or in part.

- 73C/16 (part map with area of 313 km^2)
- 73D/13 (almost full map with area of 600 km^2)
- 73D/14 (part map with area of 264 km^2)
- 73G/4 (almost full map with area of 648 km^2)
- 73H/1 (full map with area of 703.84 km^2)
- 73H/2 (almost full map with area of 621.39 km^2)
- 73G/8 (part map with area of 510 km^2)
- 73H/5 (full map with area of 703.84 km^2)
- 73H/6 (part map with area of 469 km^2)
- 73G/12 (part map with area of 20 km^2)
- 73H/9 (part map with area of 156.50 km^2) and
- 73H/10 (part map with area of 19 km^2).

All the toposheets were fed to computer and geo-coded using ERDAS 9.1 Image Processing Software. After the geo-coding, the digital toposheets were created which were mosaiced under the same projection (Polyconic/Mount Everest). A 40 km radius circle was drawn keeping its centre at Tentulia Village which was designated as the epicentre of the study.

Similarly, the satellite image obtained from NRSA, Hyderabad, was first enhanced through certain digital image processing techniques to match resolution, geo-coded and geo-rectified with the existing Survey of India toposheets. The 12 images were then merged under the same projection. The merged toposheet and the satellite images were overlapped, and a series of thematic map and calculation were generated by on-screen digitization.

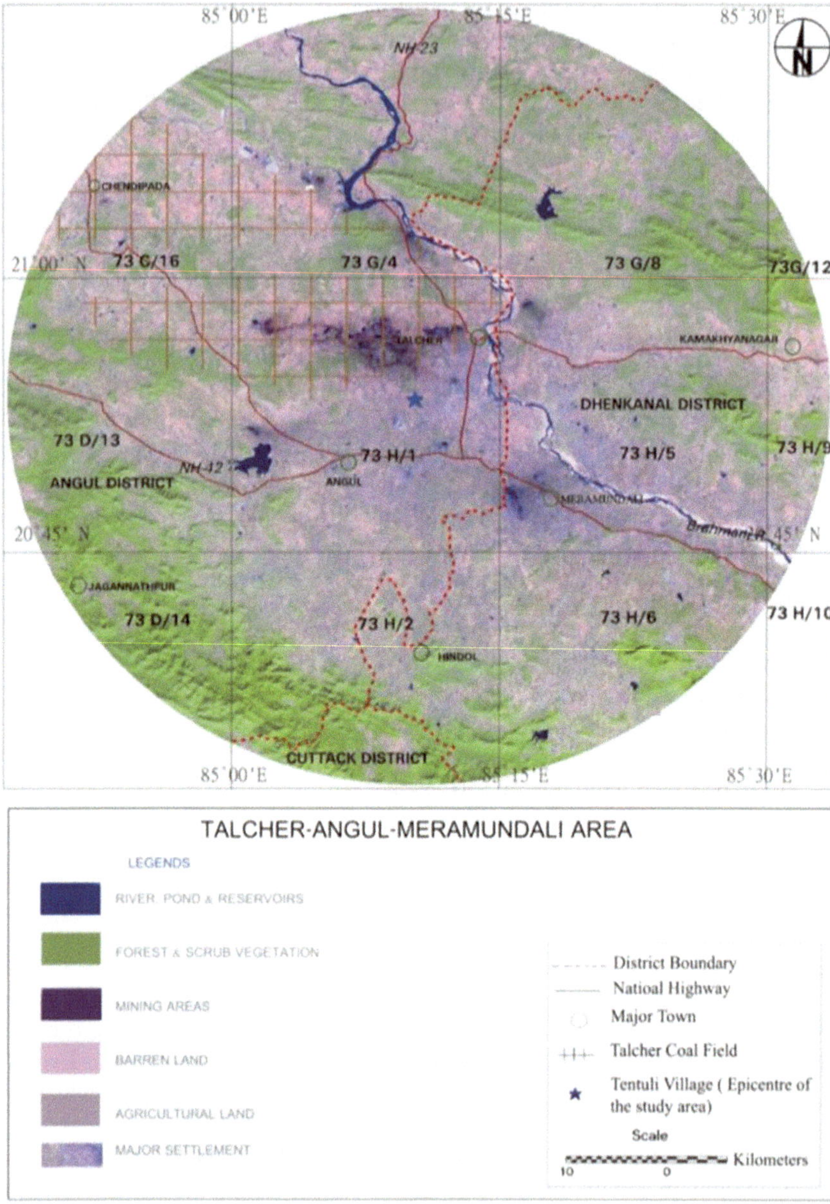

Fig. 2 Land-use of the study area as per 2007 satellite imageries

3.3 Land-Use Classification

A generalized land-use classification was adopted based on the land-use classification system developed by NRSA (1989). Similar land-use types were grouped into one generalized land-use type to form the following six main land-use types in the study area.

(1) Mining area consists of both quarries and overburden dumps.
(2) Settlements consist of built-up land, industries, roads and railway lines.
(3) Water bodies consist of rivers, streams, nalas, wetlands, reservoirs and ponds.
(4) Forest land consists of dense forest, open forest, scrub forest, forest plantation and shifting cultivation.
(5) Plantation consists of orchards and mine reclamation areas.
(6) Agriculture land consists of kharif, rabi, double crop and all types of land capable of doing agriculture, either presently cultivated or fallow land.
(7) Barren land/wasteland consists of gullied, barren rock, land with scrub and land without scrub but excluding mining area (mining area has been considered as a separate land-use unit in the study area).

Ground truth verification of prepared land-use map from remote sensing data was done to verify the interpreted land-use features along random traverses wherever approachable and feasible. The ground verification was performed by visual inspection of the interpreted land-use features in the field and also by using global positioning system (GPS). The past land-use/land-cover was deduced from the Survey of India topographic maps (1973–81). The land-use pattern of the study area is shown in Fig. 3, as per the 2007 satellite image (Singh et al. 2009).

3.4 Area Estimation of Land-Use Units

The area in square kilometre of the seven land-use units of the recent (Dec., 2007) and past (1973) land-use maps were estimated grid-wise and district-wise for the whole of the study area. This was done by using the measurement tool available with ERDAS software package. The comparison was done between the satellite image and the available toposheets.

3.5 Identification of the Barren Lands Using GIS

Barren lands are spread all over the study area, but most of this does not fulfil the industrial and environmental requirements for setting up the new industries. The use of GIS-based study has helped in identifying the important location for setting up

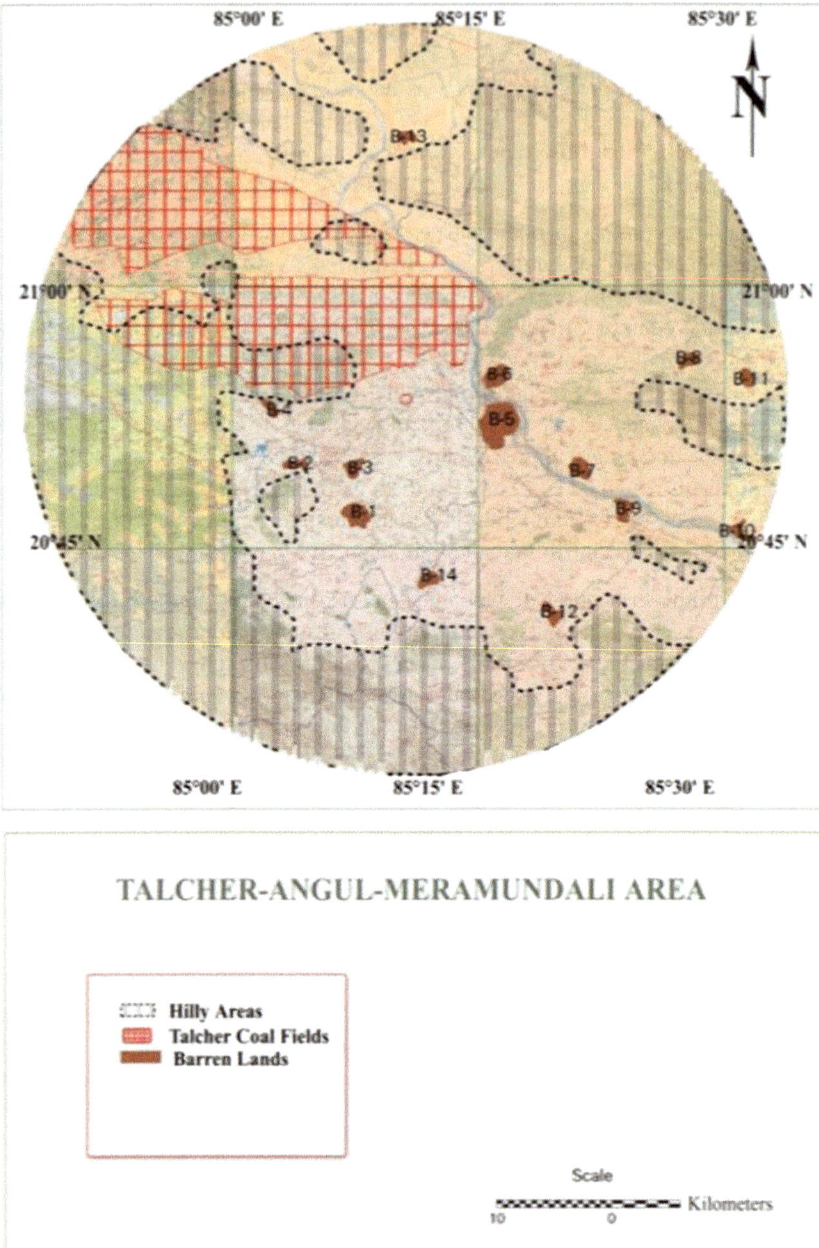

Fig. 3 Suitable locations (barren lands) for setting up industries

the locations for the new industries. Some general criteria were made for deciding of this type of barren land as listed below:

1. The barren land should not come over the hilly area which exists almost one-third of the study area.
2. This land should not come under forest lands.
3. The land should be easily accessible and closer to water sources (e.g. rivers, streams, reservoirs).
4. This should not come between the densely built-up areas.
5. The land should not come under the Talcher coalfield area.
6. Good communication network should be available near and around the proposed sites.
7. The proposed site should have an area over 2 km^2.

These criteria were fed to the computer, and using the GIS analysis, a series of tentative maps were created. This was done on the basis of overlapping of the following maps—the elevation profile of the study area, settlement, forest cover from the satellite image, the road and railway networks, access to water (distance from reservoirs, Brahmani River). Some possible 14 zones with an area of greater than 2 km^2 were identified during this process, and a final map is created and shown in Fig. 3. The possible locations were subjected to ground verifications resulting in the final maps.

The data presented in Fig. 3 reveal that agricultural land (42.31%) is the dominant land-use type in the total study area and is followed by forest land (35.2%) and barren/wasteland (14.13%). In Cuttack district, the dominant land-use is forest. While comparing the distribution of land-use types between 1973 and 2007, it was found that there is a marked increase in mining area (from 0.014 to 0.6%), barren/wasteland (0.49–14.13%) and settlements (2.12—3.8%). Area under plantation has almost doubled since 1973 (from 0.56 to 1.09%). Though there is considerable decrease in forest land (from 46.03 to 35.2%) and agricultural land (from 47.9 to 42.3%), water bodies have marginally decreased since 1973 (from 2.86 to 2.84%). Apart from the change detection in the study area, 14 places were identified using GIS studies fulfilling all the required characteristics for setting up any industries in the barren/wasteland area. Ground verification was also done to finalize these recommended barren lands for future developmental activities/industries. Requirement for barren lands for setting up these power plants keeping all environmental, economic and sociological needs is a good BMP for maintaining sustainable development in the region.

4 Conclusion

Environmental issues as a consequence of mining are dealt with respect to air quality greenhouse gases, acid rain, ground-level ozone, noise and vibrations; disturbance in hydrologic regime, acid mine drainage, land damage, land-use, subsidence, ecological restoration, social aspects are reviewed in this paper. Best management practices (BMPs) have been provided with three specific examples of societal development in coal sector: augmentation of pumped out mine water for sustainable development of coal mining regions and development planning with specific reference to land-use and land-cover in Talcher–Angul coal mining areas. These practices demonstrate that the negative impact of mining can be effectively managed by careful planning and implementation of BMPs. Further, adoption of BMPs in mining established lots of profitability decides bringing in social and environmental credibility.

References

Banejee, S.P., Singh, G.: Problems of water pollution in mining areas—issues related to its protection and control. In: Proceedings of National Seminar on Environmental Policy Issues, organised by MGMl, Kolkata (1993)

Bansal, S., Arghode V.: Mine Matters. Wastelands News (2004)

Bhattacharya, B.C.: Resettlement and rehabilitation of project affected persons (PAPs) in coal India under CSESMP. In: Mining in the 21st Century Quo Vadis? 19th World Mining Congress, New Delhi, pp 1413–1423 (2003)

BP Statistical Review of World Energy. http://www.bp.com (2005)

Dutta, M., Ghosh, R., Singh, G.: Impact of mining on water regime & its management in Jharia Coalfield, India. In: Proceedings of International Conference on 'Hydrology and Watershed Management' with a Focal Theme on Water Quality and Conservation for Sustainable Development, pp. 18–20 (2002)

GOI's Hydrocarbons Vision-2025 Report

Ghosh, R., Singh, G.: Environmental protection through water resource management in Jharia Coalfield, Jharkhand (2003)

Krishnamurthy, K.V.: Environmental impacts of coal mining in India. In: National Seminar on Environmental Engineering with Special Emphasis on Mining Environment (NSEEME) (2004)

Ministry of Environment and Forests, Government of India: Best Management Practices Manual (2003)

Ministry of Mines: Annual Report (2016)

Saxena, N.C.,Ghosh, R.: Environmental issues in coal mining-a case study. In: 5th National Symposium on Environment (1996)

Saxena, N.C., Pal, A.K.: Societal cost of environmental pollution. Minetech **21**(1), 51–54 (2000)

Saxena, N.C., Singh, G., Pathak, P., Sarkar, B.C., Pal, A.K.: Mining Environment Management Manual. Scientific Publishers, India (2005)

Singh, G.: Impact of coal mining on mine water quality. Int. J. Mine Water **7**(3), 49–59 (1988)

Singh, G.: Water pollution issues and control strategies in mining areas. Minetech **20**(1), 45–53 (1999)

Singh, G.: Regional environmental and social challenges facing the coal industry. In: The Coal Summit, Organised the India Energy Forum, Indian Coal Forum and MGMI (2005a)

Singh, G.: Water sustainability through augmentation of underground pumped out water for potable purpose from coal mines of eastern India. Environ. Geochem. **8**(1&2), 89–94 (2005b)
Singh, G., Gupta, R.K.: Water pollution from coal washeries and its impact on Damodar River. Indian Min. Eng. J. **44**(3) (2005)
Singh, G., Puri, S.K.: Air quality assessment in Korba Coalfield. Indian J. Air Pollut. Control **IV** (2), 31–41 (2004)
Singh, A.N., Raghubanshi, A.S., Singh, J.S.: Plantations as a tool for mine spoil restoration. Curr. Sci. **82**, 12 (2002)
Singh, G., Singh, P.K., Gupta, R.D.: Proceedings of the International Conference on Mineral Industry in the New Economy: Challenges and Opportunities, ISMA, Kolkata (2009)

Websites Listed

http://www.iea.org/textbase/papers/2003/ciab_sustain.pdf
http://www.atse.org.au/index.php?sectionid=1
http://www.iied.org/mmsd/mmsd_pdfs/058_downing.pdf
http://www.nswmin.com.au/minerals/coal-sustainable-future.pdf
http://www.undp.org/tcdc/bestprac/scitech/cases/st5ind.htm
http://www.economictimes.indiatimes.com>industry>ind/goods/svs>metals&mining
https://ibm.gov.in/index.php?c=pages&m=index&id=500

Numerical Evaluation of Incremental Visual Impact

V. Dentoni, B. Grosso, G. Massacci, M. Cigagna and C. Levanti

1 Introduction

The Environmental Impact Assessment (EIA) is an environmental management tool that has been used in many parts of the world since 1970 (Landscape Institute 2018). In the European Union, the EIA procedure has been enforced by the Council Directive 85/337/EEC, amended by Council/Directive 2011/92/EC. Since its introduction, EIA has become a very important tool to predict and evaluate a wide range of effects on the environment due to development and land use change. Within the overall framework of EIA operates the Landscape and Visual Impact Assessment (LVIA), a separate but closely linked process that specifically aims to ensure that all possible effects of change, both on the landscape and on the visual perception of potential observers, are taken into account in decision-making. For some environmental topics covered by EIA, it is possible to use agreed measurable variables, technical guidelines or legislative standards against which to assess the effects of a new project (Landscape Institute 2018). Assessing landscape and visual effects is clearly more difficult, as there are some aspects of landscape modification that could be objectively measured others for which the evaluation relies on qualitative and subjective judgments. In fact, LVIA seems to depend more on the subjective judgement of the observers, which involves cultural and social issues, individual perceptions, aesthetic tastes and visual comprehension, and less on the

V. Dentoni (✉) · B. Grosso · G. Massacci
DICAAR—Department of Civil and Environmental Engineering and Architecture, University of Cagliari, Via Marengo 2, Cagliari, Italy
e-mail: vdentoni@unica.it

M. Cigagna · C. Levanti
CINIGEO - Consorzio Interuniversitario Nazionale Per l'Ingegneria delle Georisorse, Rome, Italy

E. Widzyk-Capehart et al. (eds.), *Proceedings of the 18th Symposium on Environmental Issues and Waste Management in Energy and Mineral Production*,
https://doi.org/10.1007/978-3-319-99903-6_9

objective characteristics of the visible change (Nicholson 1995; Dentoni and Massacci 2013). To deal with the complexity of LVIA and support decision-making, a recent research has suggested the application of the Fuzzy Cognitive Mapping (FCM) (Misthos et al. 2017). Previous studies regarding new industrial installations, such as overhead lines, transmission towers, wind turbines, chimneys and oil refinery tanks, have considered different elements in the visual impact evaluation, such as size, design and colour of the object introduced in the landscape, landscape quality, type and number of observers (Bishop 1997, 2003).

In order to assess the effect of landscape modification, the scientific literature suggests the identification of the viewpoints from which the landscape change is visible and the subsequent estimation of the visual effect on the base of direct or indirect criteria (Shafer 1969; Briggs and France 1980). Direct methods consist in checking the landscape quality by means of customary investigations regarding the subjective perception of potential observers, while indirect methods attempt to derive the landscape quality indirectly by measuring a number of landscape components. The current prevailing tendency is based on the application of psychophysical models, where direct and indirect criteria are integrated. The latest approach allows the definition of a regression model between the judgement values expressed by a sample of observers and a number of objective variables measured on the landscape (Shafer et al. 1969). An indirect approach has been suggested in 2004 by the authors of the present article to assess the visual impact arising from surface mining and quarrying (Dentoni et al. 2004). The proposed indirect method is based in fact on the calculation of the indicator Lvi (Level of visual impact), which accounts for the two objective variables that characterise the landscape modification: the extent of the excavation visible from a given viewpoint and the chromatic contrast between the bare rock and the surrounding natural landscape. The values of the two physical parameters derive from the elaboration of digital images, taken from the most significant viewpoints. The correlation between the Lvi indicator and the subjective perception expressed by two groups of interviewees (representing the potential observers) has proved to be highly satisfactory (Dentoni and Massacci 2007).

The present article proposes an integrated procedure that includes the preliminary identification of the viewpoints from which the landscape change is visible within the impact territory (by means of the Intervisibility Analysis), the selection of the sensitive viewpoints from which the photographs need to be taken and the subsequent calculation of the Level of visual impact (Lvi). The Intervisibility Analysis (IA) can be performed by using the Digital Model Surface (DMS) of the impact area under consideration and consists in the identification of the Lines of Sight (LoS) between the observed target and all points of view within the impact territory. Having identified the viewpoints from which the altered area is visible, a selection of the sensitive points need to be performed prior to the application of the Lvi criteria. The article discusses the application of the proposed procedure to the expansion of an existing BRDA (Bauxite Residue Disposal Area) located in the south-west of Sardinia, for which the EIA procedure requires the assessment of the incremental visual impact deriving from the comparison between the ante- and post-operam conditions.

2 The Case Study

As mentioned previously, the case study to which the proposed assessment procedure has been applied is a BRDA (Bauxite Residue Disposal Area) located in the south-west of Sardinia, within the industrial area of Portovesme. The red mud basin under consideration currently occupies an area of 159 ha, within the red boundaries in Fig. 1. A recent project for the conversion of the disposal method, from lagooning to dry disposal, contemplates the basin areal expansion up to 178 ha and a vertical growth from the actual height of 26.5 m up to 46.0 m. The change in the basin dimension and the consequent effect on visual perception is the object of the Landscape and Visual Impact Assessment (LVIA) carried out within the framework of the ongoing EIA procedure. As for the others environmental effects arising from the proposed expansion project, the VIA procedure requires the evaluation of the incremental impact deriving from the comparison between the ante- and post-operam conditions.

Fig. 1 Location of the BR basin and hypothesis of future expansion

The integrated assessment procedure discussed in this article includes the identification of the significant viewpoints from which the basin is visible, before and after the basin expansion, and the calculation of the ante- and post-operam Level of visual impacts (Lvi) for each sensitive viewpoint.

3 Intervisibility Analysis

The Intervisibility Analysis allows the identification of the points from which the object under investigation (target) is entirely or partially visible. It is based on the description of the territory morphology by means of a Digital Model Surface (DMS) and the subsequent definition of the Lines of Sight (LoS) between the pixels representing the impact territory and the investigated target. For a given vertical section, Fig. 2 represents with a green profile the points from which the target is visible (i.e. interposed objects do not interrupt the LoS) and with a red profile the points from which the target is not visible (i.e. obstructing objects interrupt the LoS).

The Digital Surface Model (DSM) of Sardinia has been used to describe the morphology of the impact territory under consideration. The DSM is provided in a raster format of 1-m mesh in the native reference system WGS84/UTM32 N—EPSG: 32632 and describes the regional territory in its complexity, including the topographic surface and the artefacts (Sardegna Geoportale 2018).

The basin geometry was integrated into the DSM model through 18 target points, representing the outer edge of the summit surface as described in the conversion project, while the observer was placed at 1.7 m (standard human height) from the surface model.

The Intervisibility Analysis was carried out in a Geographic Resources Analysis Support System (GRASS) environment, using the viewshed raster image analysis algorithm, which computes the LoS joining each pixel of the DSM with the 18 target points, within a radius of 10 km from the basin barycentre. If interposed objects do not interrupt the LoS, a visibility index of 1 is associated to the pixel;

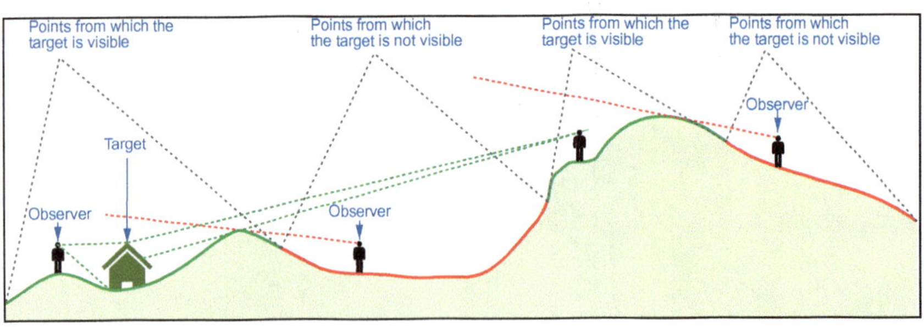

Fig. 2 Scheme of spatial intervisibility

if the LoS is interrupted, the visibility index is 0. The result for each target point consists of a raster image in which the points representing the impact territory are white (intervisibility index = 0; i.e. the target is not visible) or black (intervisibility index = 1; i.e. the target is visible).

The images relative to the target points are superimposed, adding the relative visibility index values for each pixel and normalising the result with respect to the number of target points (overall intervisibility index). The IA result for the project configuration of the basin is shown in Fig. 3. The chromatic scale indicates the visibility of the basin: dark blue (overall index = 1) represents points from which all the target points are visible; white (overall index = 0) represents points from which no target point is visible. The analysis for the post-operam condition showed that the points from which the basin is visible (overall index > 0) are mainly located within the industrial area of Portovesme and in the sea region (San Pietro canal). The points of the territory in intervisibility relationship with the basin represent 41.3% of the points within 10 km radius from the basin barycentre; 74.3% of those points are in the San Pietro canal. The IA for the ante-operam condition was performed in the same way, by inserting in DSM model the 18 target points representing the outer edge of the basin summit in the current configuration. The analysis shows that the BRDA is presently visible from 40.8% of the points

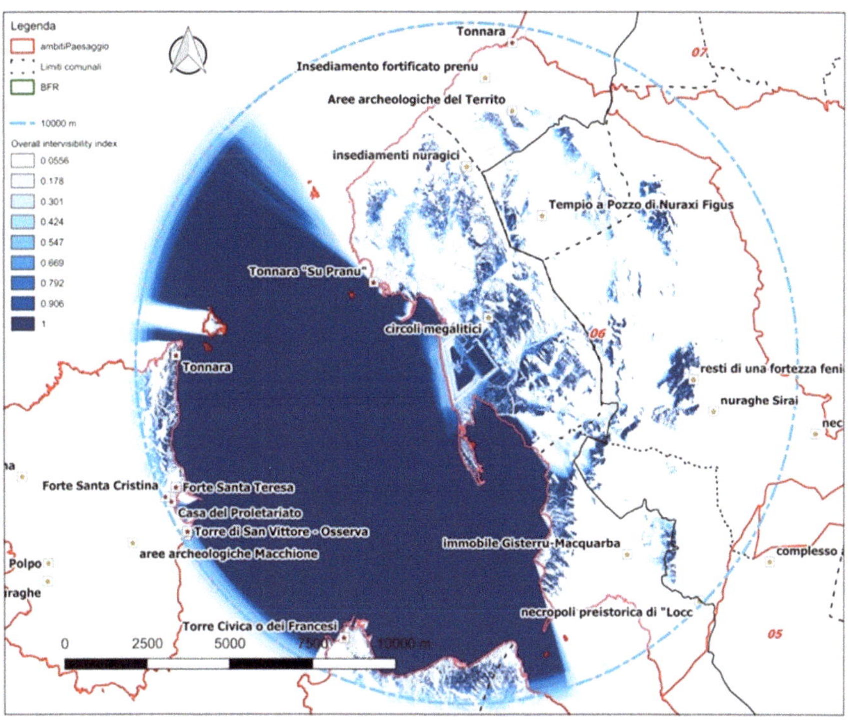

Fig. 3 Post-operam intervisibility map

included in a 10 km radius from the basin barycentre; 75.2% of those points are concentrated in the San Pietro canal. The increase in the intervisibility index due to the proposed conversion project was calculated by subtracting the two intervisibility maps (post- and ante-operam). The estimated incremental impact is about 0.5%, roughly corresponding to 2 km^2 over a total area of 406 km^2.

The selection of the sensitive viewpoints among those identified through the IA analysis was made on the basis of the following conditions: distances between 2.5 and 10 km from the basin centre; overall intervisibility index > 0; free access to people (public sites); and inclusion within scenic routes, tourist places or recreational areas.

Four sensitive viewpoints were identified: PV1 (Torre di Portoscuso, at 3.5 km from the basin); PV2 (Porto di Calasetta, at 8.5 km from the basin); PV3 (Tonnara La Punta, at 8.8 km from the basin); and PV4 (Torre di Calasetta, at 8.9 km from the basin).

4 Lvi Method

The application of the Lvi criteria allows the calculation of a single value that accounts both for the extent of the visible alteration and the chromatic contrast with the surrounding landscape. For each point of view, the Lvi criteria can be applied twice: to the photographic representation of the actual state (ante-operam) and to the photographic simulation of the proposed project (post-operam). The comparison of the two results represents a numerical indication of the incremental visual impact for a given point of view. The indicator Lvi is expressed by Eq. (1):

$$\mathrm{Lvi} = 10\log(\Omega_v/\Omega_0) + 10\log(\Delta E\mu/\Delta E_{\mathrm{BW}}) \tag{1}$$

where Ω_v is the vision solid angle subtended by the altered area from the viewpoint and Ω_0 is the threshold of visibility in a black and white colour space (8.46 · 10-8 sr);

$\Delta E\mu$ is the mean value of the Euclidean distance from the mean uniform colour of the investigated object and the colour of each pixel included in the chosen comparison surface; ΔE_{BW} is Euclidean distance between black and white (100 in the CIE Lab colour space).

Equation (2) represents the visibility level when considering only the dimension of the landscape modification:

$$\mathrm{Lv} = 10\log(\Omega_v/\Omega_0) \tag{2}$$

When using digital images, the vision solid angle subtended by the altered area can be expressed as a function of the solid angle Ω_p subtended by the entire picture:

$$\Omega_v = \Omega_p \cdot N_a/N_p \tag{3}$$

where N_{a} and N_{p}, respectively, represent the number of pixels included in the altered area and the total number of pixels in the picture.

Figure 4 shows the pictures representing the BRDA before and after the proposed enlargement, taken from the selected sensitive viewpoints listed in the previous paragraph.

A 50-mm digital camera was used to take the four photographs representing the ante-operam state, as it better reproduces the human visual field. The post-operam

Fig. 4 Ante- and images from the four sensitive viewpoints of the ante and post-operam states

condition was simulated by inserting in the same pictures the basin three-dimensional model in its final configuration. The comparison of the two scenarios (ante and post) is represented in Fig. 4 (ante-operam on the left and post-operam on the right) and shows that the BRDA expansion is not easily perceptible from the four viewpoints, given to the ratio between the size of the object and the distance from the potential observer. On the base of the photograph simulation, it seems reasonable to state that the project under examination does not cause appreciable changes in the perceptive, scenic or panoramic value of the territory.

In order to obtain a numerical indication of the incremental visual impact, the Lvi method has been applied to each pair of pictures in Fig. 4. The results are reported in Table 1, where also the Lv values are indicated to consider the effect of the basin geometric growth, regardless of the chromatic contrast with the surrounding landscape. It is worth noting that PV1 (Torre di Portoscuso) is the only point from which a change in the skyline is discernible. In order to calculate the PO (post-operam) impact from PV1, the basin geometry has been divided into two parts: beneath (PO_1) and above (PO_2) the skyline defined by the mountains behind the BRDA. A comparison surface representing the mountains beside the BFRA was selected to calculate the chromatic contrast for PO_1, while a maximum value of $\Delta E\mu$ ($\Delta E\mu = \Delta E_{BW} = 100$) was assumed for PO_1. The resulting Lvi value for PV1 (PO) is the logarithmic combination of PO_1 and PO_2.

The proposed assessment method gives a numerical indication of the visual impact increment with reference to specific shooting conditions: time and the brightness of the day, season of the year, camera's characteristics, etc. Nonetheless, the results in Table 2 are extremely significant in comparative terms, as they show that the increase in the Level of visual impact is between 3 and 6%, when considering only the basin's geometry, between 5 and 12% if the chromatic contrast is included in the impact indicator. As regards the Lvi absolute values, a previous research proved that Lvi levels lesser than 45 dB are associated with moderate visual effects (Dentoni and Massacci 2007): that level is never exceeded in the case study under consideration.

Table 1 Visual impact parameter and visual impact indicators (Lv and Lvi)

Viewpoints	Ante- or post-operam	Ω_p	N_a/N_p (%)	Ω_v (sr)	Lv (dB)	$\Delta E\mu$	Lvi (dB)
PV1	AO	0.5870	0.43	0.0025	44.7	0.1990	37.7
PV1	PO_1	0.5870	0.67	0.0039	46.7	0.2020	39.7
PV1	PO_2	0.5870	0.11	0.0006	38.7	1.0000	38.7
PV1	PO	–	0.78	0.0046	47.3	–	42.3
PV2	AO	0.1419	0.40	0.0006	38.2	0.1000	28.2
PV2	PO	0.1419	0.55	0.0008	39.6	0.1000	29.6
PV3	AO	0.3259	0.16	0.0005	37.9	0.0790	26.8
PV3	PO	0.3259	0.23	0.0008	39.5	0.0820	28.7
PV4	AO	0.1165	0.51	0.0006	38.5	0.0670	26.7
PV4	PO	0.1165	0.68	0.0008	39.7	0.0680	28.0

Table 2 Incremental visual impact (ante- and post-operam)

	PV1			PV2			PV3			PV4		
	AO	PO	Δ (%)	AO	PO	Δ (%)	AO	PO	Δ (%)	AO	PO	Δ (%)
Lv (dB)	44.7	47.3	6	38.2	39.6	4	37.9	39.5	4	38.5	39.7	3
Lvi (dB)	37.7	42.3	12	28.2	29.6	5	26.8	28.7	7	26.7	28.0	5

5 Conclusion

The present article suggests an integrated procedure for the assessment of the landscape modification visual effects, which includes the preliminary identification of the viewpoints from which the landscape change is visible (*Intervisibility Analysis*), the selection of the sensitive viewpoints and the subsequent calculation of the *Level of visual impact* (Lvi). The VIA procedure is applied to the conversion project of an existing BRDA located in the south-west of Sardinia, for which the EIA procedure requires the assessment of the incremental visual impact deriving from the comparison between the ante and post-operam states.

The increase in the *intervisibility index* due to the proposed conversion project was calculated by subtracting the intervisibility maps of the post and ante-operam conditions, respectively; the estimated incremental impact was about 0.5%, roughly corresponding to 2 km^2 over a total area of 406 km^2.

As regards the application of the Lvi method, four sensitive viewpoints were selected among those identified through the Intervisibility Analysis. The increase in the Level of visual impact (Lvi) was found to be between 5 and 12%, with a maximum value of 42.3 dB: previous researches proved that Lvi values lesser than 45 dB are associated with moderate visual effects.

Possible developments of the proposed procedure might include the definition of numerical weighting factors to integrate the Lvi equation and account for the quality of the impact territory under examination. This way the Lvi value calculated for different cases could be referred to agreed thresholds of visual impact, irrespectively of the specific site characters.

Acknowledgements Investigation carried out in the framework of projects conducted by CINIGeo (National Inter-university Consortium for Georesources Engineering, Rome, Italy). This work is part of the research project "RE-MINE -REstoration and remediation of abandoned MINE sites", funded by the Fondazione di Sardegna and Regional Sardinian Government (Grant CUP F72F16003160002).

References

Bishop, I.D.: Testing perceived landscape colour difference using the internet. Landsc. Urban Plan. **37**, 187–196 (1997)

Bishop, I.D.: Assessment of visual qualities, impacts, behaviours, in the landscape, by using measures of visibility. Environ. Plan. **30**, 677–688 (2003)

Briggs, D.J., France, J.: Landscape evaluation: a comparative study. J. Environ. Manage. **10**, 263–275 (1980)
Dentoni, V., Massacci, G.: Assessment of visual impact induced by surface mining with reference to a case study located in Sardinia (Italy). Environ. Earth Sci. **68**, 1485–1493 (2013)
Dentoni, V., Massacci, G., Meloni, R.: Landscape and Visual Impact Assessment of Opencast Mining. In: Proceedings of the 8th International Symposium on Environmental Issues and Waste Management in Energy and Mineral Production (2004), Antalya, Turkey, 17th–20th May 2004: 113–118. ISBN 975-6707-11-9
Dentoni, V., Massacci, G.: Visibility of surface mining and impact perception. Int. J. Min., Reclam. Environ. **21**(1), 6–13 (2007). Taylor & Francis, ISSN 1748-0930 print, ISSN 1748-0949 online
Directive 2011/92/EU of the European Parliament and of the Council of 13 December 2011 on the assessment of the effects of certain public and private projects on the environment
Landscape Institute: Guidelines for landscape and visual impact assessment, at https://www.landscapeinstitute.org/. Last accessed 17 May 2018
Misthos, L.M., Messaris, G., Menegaki, M., Damigos, D.: Exploring the perceived intrusion of mining into the landscape using the fuzzy cognitive mapping approach. Ecol. Eng. **101**, 60–74 (2017)
Nicholson, D.T.: The visual impact of quarrying. Quarry Manag. **22**(7), 39–42 (1995)
Sardegna Geoportale, http://www.sardegnageoportale.it/navigatori/sardegnamappe/. Last accessed: 17 May 2018
Shafer, E.L.: Perception of natural environments. Environ. Behav. **1**, 71–82 (1969)
Shafer, E.L., Hamilton, J.F., Schmidt, E.A.: Natural landscape preferences: a predictive model. J. Leisure Res. **1**, 1–19 (1969)

Possible Environmental Risks Associated with Steel Slag: A Batch Study

S. Biliangadi, V. N. L. Wong, M. Yellishetty, A. Kumar Dikshit and S. Majumdar

1 Introduction

The management of mining, power plant, and metallurgical wastes/by-products is of a great challenge as well as an opportunity to create something good out of it. Quantity wise numbers are surging every year. Along with regulation of these outputs as environmental concerns are growing, push toward recycling and reusing is also gaining popularity—moreover, it is a need of the hour. Overburden, fly ash, iron slag, steel slag, and many such materials are used by different sectors such as construction, cement, roads, and agriculture. A highly alkaline, porous, crystalline, heterogeneous, and non-hazardous (Matei 2007; Proctor et al. 2000; Ziemkiewicz 1998) by-product from steelmaking plant is basic oxygen furnace (BOF)/steel/ Linz-Donawitz (LD) slag. Every tonne of steel generates 150–200 kg of steel slag, and most of it ends up in storage yards due to low market value and demand and high hauling costs (Drissen et al. 2009; Kumar et al. 2010). Expansive nature of slag due to free lime restricts its usage in cement and construction industries.

S. Biliangadi (✉)
IIT B-Monash Research Academy, IIT Bombay, Mumbai 400076, India
e-mail: shashibhushanbiliangadi@gmail.com

V. N. L. Wong · M. Yellishetty
Monash University, Clayton, VIC 3800, Australia

A. K. Dikshit
IIT Bombay, Mumbai 400076, India

S. Majumdar
JSW Group, Mumbai, India

E. Widzyk-Capehart et al. (eds.), *Proceedings of the 18th Symposium on Environmental Issues and Waste Management in Energy and Mineral Production*,
https://doi.org/10.1007/978-3-319-99903-6_10

Synthetic soil or artificial soil is one such possibility where by-products from different industries are blended together to create a topsoil (Yellishetty et al. 2014). The main purpose of this topsoil would be to replace the natural soil with synthetic (top) soil for rehabilitation and restoration works such as: for mine rehabilitation, landfill cover, road median, and public gardens. Slag, when used as soil, would get exposed to different environmental conditions imparted mainly by water, organic matter or humic substances, acidic wastes/saline wastes.

Organic waste compost in an amendment made to slag to build synthetic (top) soil. Compost provides carbon, nitrogen, microbes as well as readily available inorganic ions such as Fe, Zn, K, Na, and others. Compost increases water holding capacity, provides above-mentioned nutrients, and enhances soil structure. Also, high concentrations of Na and K in compost would result in high conductivity. Environmentally, harmful trace metals such as Pb, Zn, Cr, and others are found in soluble phase under acidic conditions. Organic acids and humic acids present/generated would develop acidic conditions as well. Hence, the right approach would be to study the behavior of slag under acidic and saline conditions. In the current study, these environmental conditions are simulated using 0.11 M acetic acid and a mixture of 0.128 M sodium dithionite, 0.3 M sodium citrate, and 0.1 M sodium bicarbonate solutions (Shivpuri et al. 2012). The main objective is to measure the amount of inorganic ions extracted under acidic and saline conditions.

2 Materials

Linz-Donawitz (LD) slag was supplied by JSW Vijaynagar plant, Toranagallu, Karnataka, India. Two extraction fluids (1) 0.11 M acetic acid and (2) a mixture of 0.128 M sodium dithionite, 0.3 M sodium citrate, and 0.1 M sodium bicarbonate extraction solutions were prepared using analytical grade reagents. Milli-Q water was used for dilution.

3 Experimental Methodology

A series of nine 100 mL centrifuge tubes having 0.5 g (size < 150 μ) fresh LD slag was filled with 50 mL of extraction fluid (Shivpuri 2012). Tubes were placed on the end-over-end mixer (Make: Trishul Equipments, Mumbai). Sampling was done at regular intervals such as 1, 2, 4, 8, 12, 18, 24, 30, 36, and 48 h. Extraction fluid was filtered with a glass filter, acidified with 2% nitric acid for inductively coupled plasma atomic emission spectroscopy (ICP-AES) analysis (Make: SPECTRO Analytical Instruments GmbH, Germany). All experiments were performed in duplicates.

4 Results and Discussions

Cumulative metal concentrations (mg/kg) with time extracted using 0.11 M acetic acid and a mixture of 0.128 M sodium dithionite, 0.3 M sodium citrate, and 0.1 M sodium bicarbonate extraction solution are shown in Figs. 1 and 2. The rapid desorption of metal ions present on the surfaces of porous, heterogeneous, and crystalline slag was restricted to initial five hours for most of the metals when slag got exposed to the 0.11 M acetic acid solution. It took five hours for readily available metal ions like Al, Cd, Ni, and Zn to reach equilibrium conditions. In first five hours, ions present on an outer surface, mainly macropores, started coming out of solid phase into the solution phase quickly under acidic conditions (initial pH of 0.11 M acetic acid was 2.93). However, it was not the same for ions like Ca and Pb —which reached equilibrium slowly (Fig. 1). As contact time between slag and solution increased, Ca and Pb concentrations were steadily increased throughout 48 h. Indicating desorption of Ca and Pb from meso- and micropores throughout the experiment period.

Steel slag exposure to freshly prepared Na rich solution also showed a similar kind of trend. Except for Al, Ca, and Fe, all other metal ions reached equilibrium in less than five hours of contact between slag and extraction solution. However, Al, Ca, and Fe continued to release into the solution throughout the contact period. Indicating enhanced release of ions under saline conditions (initial conductivity of solution was 60.6 mS/cm).

Table 1 describes percentage increase in metal extraction under saline and acidic conditions. Al, Ba, Cr, Cu, Fe, Mn, and Zn extraction was 680.09, 164.71, 187.41, 137.04, 2066.79, 191.01, and 281.48%, respectively, times more in saline conditions than in acidic conditions. Extraction of Pb was 17.78% less in saline conditions compared to acidic conditions. This implies care should be taken while using highly saline organic waste compost as an amendment to steel slag to build an artificial/synthetic soil. To predict the release of metals in future, long-term leaching studies should be performed with actual artificial soils.

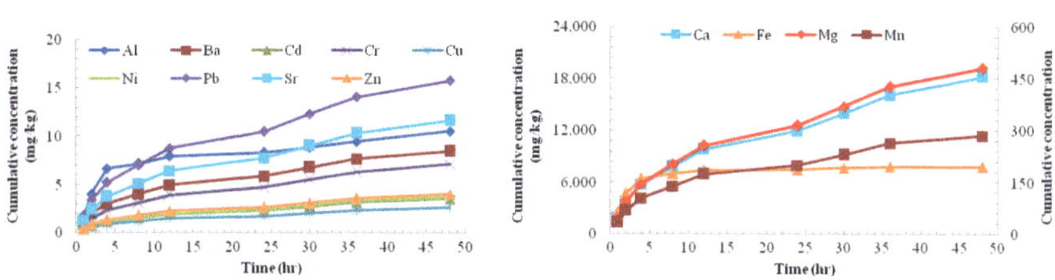

Fig. 1 Cumulative concentration of metals extracted using 0.11 M acetic acid with time

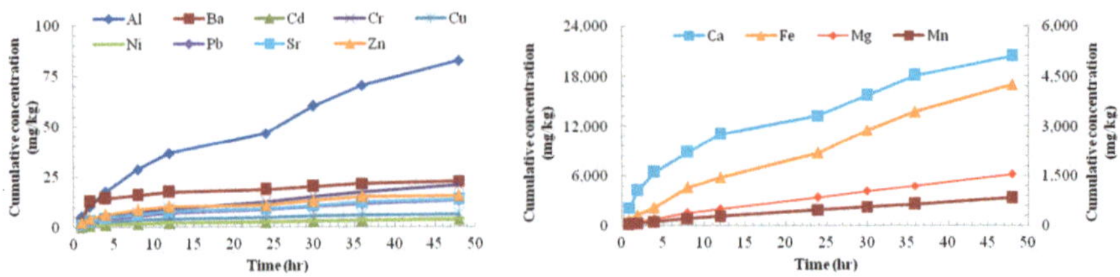

Fig. 2 Cumulative concentration of metals extracted using a mixture of 0.128 M sodium dithionite, 0.3 M sodium citrate, and 0.1 M sodium bicarbonate extraction solution with time

Table 1 Percentage increase in metal extracted at the end of 48 h

Metal	A (mg/kg)	B (mg/kg)	% increase (B-A)/A × 100
Al	10.55	82.30	680.09
Ba	8.50	22.50	164.71
Ca	18188.55	20415.95	12.25
Cd	3.60	3.60	0.00
Cr	7.15	20.55	187.41
Cu	2.70	6.40	137.04
Fe	196.50	4257.75	2066.79
Mg	480.65	1532.95	218.93
Mn	285.80	831.70	191.01
Ni	3.80	3.80	0.00
Pb	15.75	12.95	−17.78
Sr	11.65	13.55	16.31
Zn	4.05	15.45	281.48

Note A: Cumulative concentration of metal extracted using 0.11 M acetic acid after 48 h. **B**: Cumulative concentration of metal extracted using a mixture of 0.128 M sodium dithionite, 0.3 M sodium citrate, and 0.1 M sodium bicarbonate extraction solution after 48 h

5 Conclusions

The utilization of industrial wastes and municipal waste in individual or together would definitely help to reduce the burden on land and would also curb the pollution level to a certain extent. However, the change that can happen in future cannot be neglected. The outcomes of batch extraction study on steel slag can be summarized as follows:

1. Steel slag as synthetic soil would be exposed to acidic and saline conditions. The result showed that metal extraction was more with saline extraction conditions.

2. Al, Ba, Cr, Cu, Fe, Mn and Zn extraction was 680.09, 164.71, 187.41, 137.04, 2066.79, 191.01, and 281.48%, respectively, more in saline conditions than in acidic conditions.
3. Extraction of Pb was 17.78% less in saline conditions compared to acidic conditions.
4. Care should be taken while using highly saline (Na rich) compost as an amendment to alkaline steel slag to make synthetic soil, as it could enhance the trace metal leaching.

Acknowledgments We are thankful to JSW Steel for providing the LD slag from Vijaynagar works, Karnataka, India for supporting our research.

References

Drissen, P., Ehrenberg,A., Kühn, M., Mudersbach, D.: Recent development in slag treatment and dust recycling. Process Metall. 737–745 (2009)

Kumar, D.S., Umadevi, T., Paliwal, H.K., Prasad, G., Mahapatra, P.C., Ranjan, M., Steel, J.S.W.: Recycling steelmaking slags in cement. World Cem. **11**, 1–6 (2010)

Matei, E., Predescu, C., Sohaciu, M., Berbecaru, A., Independentei, S., Jf, H.: Method for assessment of potentially pollution of slag from iron and steel industry. In: Proceedings of 3rd IASME/WSEAS International Conference on Energy, Environment Ecosystems and Sustainable Development. Ag. Nikolaos, Greece, pp. 24–27 (2007)

Proctor, D.M., Fehling, K.A., Shay, E.C., Wittenborn, J.L., Green, J.J., Avent, C., Bigham, R.D., Connolly, M., Lee, B., Shepker, T.O., Zak, Ma.: Physical and chemical characteristics of blast furnace, basic oxygen furnace, and electric arc furnace steel industry slags. Environ. Sci. Technol. **34**(8), 1576–1582 (2000)

Shivpuri, K.K., Lokeshappa, B., Kulkarni, L.B., Dikshit, A.K.: Metal leaching potential in coal fly ash. Am. J. Environ. Eng. **1**, 21–27 (2012)

Yellishetty, M., Wong, V., Taylor, M., Li, J.: Forming artificial soils from waste materials for mine site rehabilitation. **16**, 16445 (2014)

Ziemkiewicz,P.: Steel slag : applications for AMD control. In: Proceedings of the 1998 Conference on Hazardous Waste Research, no. 304, pp. 44–62 (1998)

Part III
Management of Mining and Hazardous Waste and Waste Stabilization

Application of Fly Ash to Acidic Soil to Improve Plant Growth in Disturbed Land of Open-Cut Mining

A. Hamanaka, H. Yamasaki, T. Sasaoka, H. Shimada and S. Matsumoto

1 Introduction

Rehabilitation of post-mine land is necessary for sustainable development of open-cut mining. Minerals are mined with the creation of the large disturbed land in open-cut mining. Thus, rehabilitation of post-mine land is required at the end of the development for environmental conservation. In open-cut mining, topsoil, which is formed from the surface down to approximately 1.0 m depth, is stored during the excavation stage, followed by the placement in surface layer in post-mine land during rehabilitation since they contain much nutrition that is useful for plant growth (Sheoran et al. 2010). However, the shortage of topsoil attributing to the loss while hauling soil results in the difficulty of rehabilitation in some cases. Furthermore, the loss of topsoil caused by soil erosion or landslide causes the inhibition of plant growth in the post-mined land, leading to the failure of rehabilitation (Sheoran et al. 2010; Takeuchi and Shimano 2009). Therefore, it is important to secure enough amount of topsoil for rehabilitation in the disturbed land of open-cut mining.

In many mines, acid sulfate soils and/or rocks are mined as waste materials with excavation of minerals (Ueno 2004). These soils and/or rocks consist of several kinds of minerals such as silicate and sulfide minerals (e.g. pyrite: FeS_2). This sulfide mineral leads to Acid Mine Drainage (AMD) when they are exposed to oxygen and water during excavation process as follows (Nugraha et al. 2009; Nordstrom et al. 2000; Nordstrom and Alpers 1999):

A. Hamanaka (✉) · H. Yamasaki · T. Sasaoka · H. Shimada
Department of Earth Resources Engineering, Kyushu University, Fukuoka, Japan
e-mail: hamanaka@mine.kyushu-u.ac.jp

S. Matsumoto
Geological Survey of Japan, National Institute of Advanced Industrial Science and Technology (AIST), Tokyo, Japan

E. Widzyk-Capehart et al. (eds.), *Proceedings of the 18th Symposium on Environmental Issues and Waste Management in Energy and Mineral Production*,
https://doi.org/10.1007/978-3-319-99903-6_11

$$FeS_2 + 7/2O_2 + H_2O \rightarrow Fe^{2+} + 2SO_4^{2-} + 2H^+ \quad (1)$$

$$Fe^{2+} + 1/4O_2 + H^+ \rightarrow Fe^{3+} + 1/2H_2O \quad (2)$$

$$Fe^{3+} + 3H_2O \rightarrow Fe(OH)_3 + 3H^+ \quad (3)$$

$$FeS_2 + 14Fe^{3+} + 8H_2O \rightarrow 15Fe^{2+} + 2SO_4^{2-} + 16H^+ \quad (4)$$

The waste materials are classified as Potentially Acid Forming (PAF) by means of several geochemical tests: PAF is the source of AMD. PAF is, generally, backfilled under the ground to prevent the occurrence of AMD. However, they are mixed in topsoil and backfilled in surface layer in the post-mining land in some cases (Matsumoto et al. 2015). In this situation, the mixing of such waste materials consisting of PAF in topsoil causes soil acidification, and this affects revegetation process in mining operation. In fact, the failure of revegetation by soil acidification attributed to metal dissolutions with PAF was reported in Indonesian open-cut coal mines (Ogata 2016). However, another researcher reported vegetation growth can be promoted if the soil quality is good for plant growth, although the soil contains high metal contents (Anawar et al. 2013).

As for the vast amount of fly ash produced as the industrial wastes in coal-fired power plants, the effective utilization ways are expected in some applications as alternatives for natural materials. Some of fly ash is utilized such as a material for cements though the others are disposed in ash dams. Considering the preparation and maintenance costs of landfill area and environmental issues, it is very meaningful to discuss the utilization of fly ash except a cement usage. Besides, the potential to utilize fly ash for biomass production has been suggested (Pandey et al. 2009). Several studies have shown that the fly ash application on topsoil can promote the plant growth because it can increase pH of the acid soil (Capp and Adams 1971; Srivastava and Chhonkar 2000; Brown et al. 1997), improve the soil texture (Fail and Wochok 1977), and supply the trace element which is necessary for plant growth (Khan and Khan 1996). On the other hand, careful consideration is needed when the amendment of topsoil using fly ash is discussed because the excess addition of fly ash causes the decrease in plant growth due to the salinity of the soils and impact from toxic elements (Singh and Siddiqui 2003; Matsi and Keramidas 1999). Given these facts, several researchers suggested the allowable range of fly ash addition to the soil based on pot experiment (Matsi and Keramidas 1999; Mishra and Shukla 2009). However, the use of fly ash for soil amendment is a complicated topic because the properties of fly ash and acidic soils are site specific.

The present work is focused on addition of fly ash which has high alkaline to two acidic soils which have different metal leachate potential. Therefore, the utilization of fly ash for preparation of seedbed in disturbed land in open-cut mine is discussed by means of laboratory pot trials using simulated acidic soil with a mixture of pyrite, fly ash which has higher alkalinity, and *Acacia mangium* which is a typical species planted for fast growing tree in the beginning stage of revegetation.

2 Materials and Methods

2.1 Preparation of Simulated Soil for Pot Trials

Simulated acid soil was prepared by mixing decomposed granite soil and pyrite. In this study, two types of acidic soils were prepared to simulate the different potential of metal leachate from the soil. Besides, fly ash (FA) was taken in the coal-fired power plant in Japan. Table 1 shows fundamental properties of each sample. Paste pH and EC tests were carried out by the dissolution process with sample and deionized water. The mixing ratio of sample and deionized water was 1:2.5 in the paste pH (Kabala et al. 2016) and 1:5 in the paste EC (Slavich and Petterson 1993), respectively, to compare the standard (Research Committee of Japanese Institute of Landscape Architecture 2000). From the Table 1, the acidic soil 2 shows the lower pH and higher EC compared to acidic soil 1, meaning that the potential of metal leachate is higher. Fly ash has the higher alkalinity, and the dominant particle size is silt.

Paste pH and EC, dissolved metal ion analysis, and permeability test based on ASTM D5084-10 standard were conducted for ten types of samples which were prepared by changing the mixture rate of simulated acidic soil and fly ash (Tables 2 and 3). Paste pH rises with an increase in mixing ratio of fly ash in both acidic soils 1 and 2. Above all, excess addition of fly ash causes soil alkalization in acidic soil 1 because the paste pH shows alkaline when the mixture ratio of fly ash is more than 20%. Paste EC declines under the neutral pH condition when the fly ash mixture ratio is 5 and 10% for acidic soil 1 and more than 10% for acidic soil 2, respectively. This fact indicates that metal dissolution can be minimized by controlling the soil pH. However, the remediation of paste EC in acidic soil 2 is not enough due to the high potential of metal dissolution from acidic soil based on the standard (Research Committee of Japanese Institute of Landscape Architecture 2000). The hydraulic conductivity is slightly decreased with an addition of fly ash. Table 3 shows the result of metal ion analysis by using the water from paste EC test. The major metal contents eluted in AMD, such as Al, Fe, and Mn, were measured in this test. The dissolution of Al causes the inhibition of plants' growth due to the elongation of roots of plants (Kochian et al. 2005; Kikui et al. 2005). Additionally, Fe derived from the pyrite is a dominant metal ion eluted under the acidic conditions, and its concentration is higher in acidic soil 2.

Table 1 Soil compositions, paste pH, paste EC of each sample

Sample	Sand (%)	Silt (%)	Clay (%)	Soil texture	Paste pH	Paste EC (mS/cm)
Acidic soil 1	84.5	7.3	8.3	Sandy loam	3.70	1.32
Acidic soil 2	85.8	7.2	7.0	Sandy loam	3.10	3.50
Fly ash	9.0	75.3	15.7	Silt loam	12.60	3.20

Table 2 Paste pH, paste EC, hydraulic conductivity of simulated soils

	Fly ash (%)	Paste pH	Paste EC (mS/cm)	Hydraulic conductivity (cm/s)
Acidic soil 1	0	3.70	1.32	5.70×10^{-4}
	5	5.90	1.06	2.16×10^{-4}
	10	6.35	1.20	1.34×10^{-4}
	20	8.44	1.33	1.13×10^{-4}
	50	11.22	1.49	1.56×10^{-5}
Acidic soil 2	0	3.10	3.50	7.43×10^{-4}
	5	4.83	3.30	2.84×10^{-4}
	10	5.64	2.80	1.45×10^{-4}
	20	6.13	2.80	1.33×10^{-4}
	50	7.87	2.40	3.34×10^{-5}
Standard		4.5–7.5	0.1–1.0	10^{-4}–10^{-1}

Table 3 Dissolution of dominant metal ions in AMD

	Fly ash (%)	Al (ppm)	Fe (ppm)	Mn (ppm)
Acidic soil 1	0	34.1	298	14.2
	5	5.4	143	8.1
	10	15.7	161	9.3
	20	17.6	143	9.3
	50	14.1	17	0.2
Acidic soil 2	0	20.2	1067	20.7
	5	24.6	1167	23.5
	10	14.4	814	17.8
	20	31.8	705	16.2
	50	26.2	187	6.4

2.2 Laboratory Pot Trials

In order to discuss the applicability of fly ash on acidic conditions to promote the plant growth, *A. mangium* was planted on the prepared soils mixed with acidic soil and fly ash described in previous section. The plants were placed in the phytotron glass room G-9 in Biotron Application Center, Kyushu University, under the following conditions: at 30 °C room temperature and 70% relative humidity assuming the tropical climate. In this test, five plants were planted in each soil, and the height and the basal diameter were measured every a week. At the end of the test, the dry weight of *A. mangium* was measured. The pot trials continued for 11 weeks until a clear distinction was observed. All pots were supplied with 500 mL water every 3∼4 day, while liquid fertilizer HYPONeX formula 6-10-5 diluted to 1000 ppm with water was additionally added weekly. The data obtained from the laboratory pot trials are processed by using Tukey-kramer method to indicate the significant differences among each result.

3 Results and Discussion

3.1 Preparation of Simulated Acidic Soil

Figure 1a, b shows the photograph of *A. mangium* at the end of the test. We also prepared the pots of decomposed granite soil without pyrite and fly ash as a control pot. It can be clearly observed that the growth of *A. mangium* is promoted with mixing of fly ash in acidic soil 1, whereas most of the plants do not grow in acidic soil 2. Figures 2a, b and 3a, b show the increment of plant height and basal diameter. These increments are found in acidic soil 1 with 5, 10, and 20% of fly ash with elapsed time though the other soil shows the slight increase or no improvement due to the death of plant. The height increase in the plants is improved in acidic soil 1. The acidic soil 1 with 10% of fly ash addition showed the highest height

Fig. 1 Photograph of *A. mangium* at the end of test: **a** acidic soil 1, **b** acidic soil 2

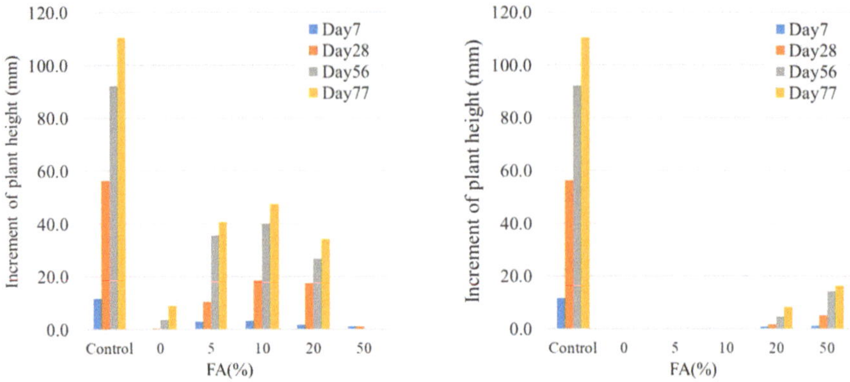

(a) Acidic soil 1 (b) Acidic soil 2

Fig. 2 Increment of plant height: **a** acidic soil 1, **b** acidic soil 2

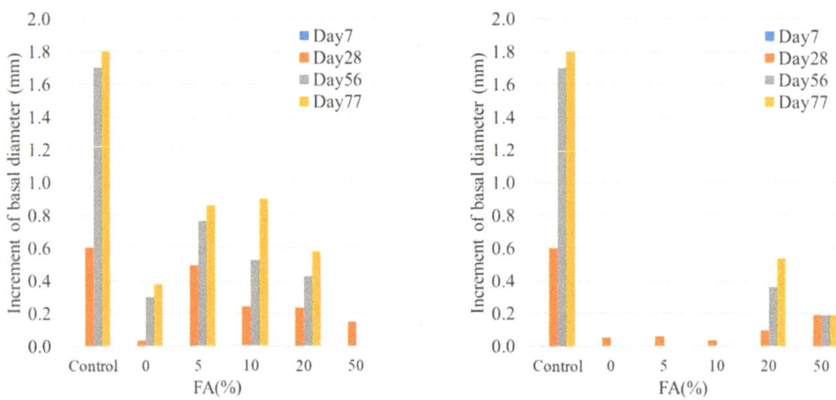

(a) Acidic soil 1 (b) Acidic soil 2

Fig. 3 Increment of basal diameter: **a** acidic soil 1, **b** acidic soil 2

(47.3 mm) followed by 5% of fly ash addition (40.6 mm) at the end of the test, while the height increase in the control pots is 110.5 mm. The basal diameter increase showed similar pattern with that of height. Maximum basal diameter was found in acidic soil 1 with 10% of fly ash (0.90 mm) followed by 5% of fly ash addition (0.86 mm) while the basal diameter of the control pots is 1.80 mm. A slight increment or no improvements of both of plant height and basal diameter are found in acidic soil 2.

Dry weight of *A. mangium* is a proper indicator of the impact of soil conditions on the plant growth in terms of biomass. Multiple comparison analysis of dry weight revealed significant differences between the pots in acidic soil 1 (Fig. 4). The dry weight is significantly increased in acidic soil 1 with 5% (0.88 g) and 10% (0.71 g) of fly ash followed by 20% of fly ash (0.39 g) while the weight of the

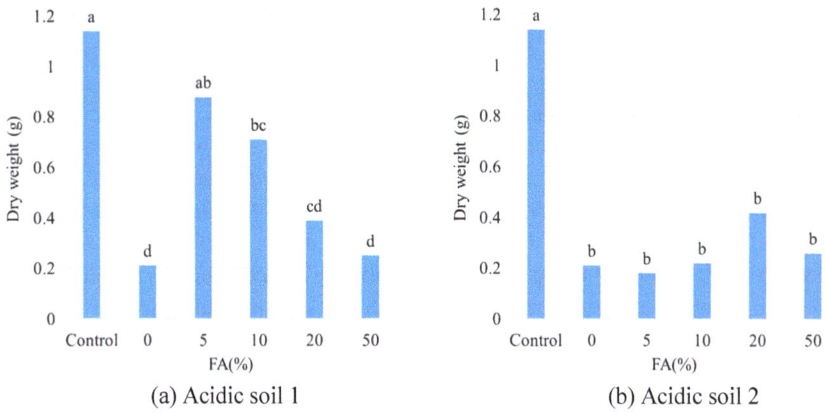

Fig. 4 Dry weight of *A. mangium*: **a** acidic soil 1, **b** acidic soil 2

control pots is 1.14 g. The 5% addition of fly ash seems to be appropriate to promote the plant growth because the results did not show statistical difference with a control. The pots of acidic soil 1 show the addition of 0 and 50% does not promote plant growth. Considering that the inhibition of the growth of plants by aluminum is a major problem in acidic conditions, the content of Al dissolved under the acidic condition may affect the growth of *A. mangium* in the 0% of fly ash. The dry weight is declined with an increase in fly ash addition under the alkaline conditions. The decline of the dry weight is probably due to salinity caused by sulfate, chloride, carbonate, bicarbonate from fly ash. Additionally, the reduction in absorbing nutrient under the alkaline condition is also a factor to inhibit the plant growth. On the other hand, the results of acidic soil 2 do not show statistically significant difference between the pots. The significant inhibitions of plant growth in acidic soil 2 are due to the osmotic stress caused by high salinity and the specific effects of individual ions. Previous research indicated water uptake by the root is reduced if there is osmotic stress (Bernstein 1975). The electric conductivity of acidic soil 2 shows quite higher compared to the standard value (Table 2), meaning that the water uptake by roots is inhibited due to the effects of osmotic stress. Besides, the specific ions should influence the plant growth because excessive metal ions dissolved. Therefore, the metal accumulation in the plant body will be investigated as the future task.

4 Conclusions

Vegetation process is necessary after mining operation in terms of environmental conservation. However, acid sulfate soils attributed to the process of backfill of waste rocks in post-mining area inhibit plant establishment in the revegetation stage in some cases. Using fly ash which has alkaline properties seems to be an effective way to neutralize acidic soils. In this research, laboratory pot trials were performed with *A. mangium* for 11 weeks in order to understand the effects of fly ash addition on plant growth in acid-topsoil. According to the results, it is possible to improve the plant growth under the acidic soil by remediating soil conditions with fly ash addition though excess of fly ash inhibits the plant growth due to the salinity and alkalization of soils. Moreover, it is difficult to improve the plant growth by using fly ash if the acidic soil has the high potential of metal dissolution. In summary, effective revegetation by using fly ash is possible to be performed by evaluating the effect of metal dissolution from acidic soil and fly ash on plant growth.

References

Anawar, H.M., Canha, N., Santa-Regina, I., Freitas, M.C.: Adaptation, tolerance, and evolution of plant species in a pyrite mine in response to contamination level and properties of mine tailings: sustainable rehabilitation. J. Soils Sed. **13**(4), 730–741 (2013)

Bernstein, L.: Effects of salinity and sodicity on plant-growth. Ann. Rev. Phytopathol. **13**, 295–312 (1975)

Brown, T.H., Bland, A.E., Wheeldon, J.M.: Pressurized fluidized bed combustion ash. 2. Soil and mine spoil amendment use options. Fuel **76**(8), 741–748 (1997)

Capp, J.P., Adams, L.M.: Reclamation of coal mine wastes and strip spoil with fly ash. In: Proceedings of Economic use of surface-mined land and mine refuse symposium, pp. 26–37 (1971)

Fail, J.L., Wochok, Z.S.: Soybean growth on fly ash-amended strip mine spoils. Plant Soil **48**(2), 473 484 (1977)

Kabala, C., Musztyfaga, E., Galka, B., Labunska, D., Manczynska, P.: Conversion of soil pH 1:2.5 KCl and 1:2.5 H2O to 1:5 H2O: conclusions for soil management, environmental monitoring, and international soil databases. Pol. J. Environ. Stud. **25**(2), 647–653 (2016)

Khan, M.R., Khan, M.W.: The effect of fly ash on plant growth and yield of tomato. Environ. Pollut. **92**(2), 105–111 (1996)

Kikui, S., Sasaki, T., Maekawa, M., Miyao, A., Hirochika, H., Matsumoto, H., Yamamoto, Y.: Physiological and genetic analyses of aluminium tolerance in rice, focusing on root growth during germination. J. Inorg. Biochem. **99**(9), 1837–1844 (2005)

Kochian, L.V., Pineros, M.A., Hoekenga, O.A.: The physiology, genetics and molecular biology of plant aluminum resistance and toxicity. Plant Soil **274**(1–2), 175–195 (2005)

Matsi, T., Keramidas, V.Z.: Fly ash application on two acid soils and its effect on soil salinity, pH, B, P and on ryegrass growth and composition. Environ. Pollut. **104**(1), 107–112 (1999)

Matsumoto, S., Ogata, S., Dwiki, S., Hideki, S., Sasaoka, T.: Fundamental research on evaluation of acid-topsoil for effective revegetation at post-mining land in indonesian open cast coal mine. Proc. Int. Sym. Earth Sci. Technol. **2015**, 65–69 (2015)

Mishra, L.C., Shukla, K.N.: Effects of fly-ash deposition on growth, metabolism and dry-matter production of maize and soybean. Environ. Pollut. Ser. Ecol. Biol. **42**(1), 1–13 (2009)

Nordstrom, D.K., Alpers, C.N.: Negative pH, efflorescent mineralogy, and consequences for environmental restoration at the iron mountain superfund site, California. Proc. Natl. Acad. Sci. USA **96**(7), 3455–3462 (1999)

Nordstrom, D.K., Alpers, C.N., Ptacek, C.J., Blowes, D.W.: Negative pH and extremely acidic mine waters from Iron mountain, California. Environ. Sci. Technol. **34**(2), 254–258 (2000)

Nugraha, C., Shimada, H., Sasaoka, T., Ichinose, M., Matsui, K., Manege, I.: Geochemistry of waste rock at dumping area. Int. J. Min. Reclam. Environ. **23**(2), 132–143 (2009)

Ogata, S., Matsumoto, S., Hideki, S., Sasaoka, T.: Study on effective revegetation at post-mining land in Indonesian open-cut coal mine. Proc. Int. Sym. Earth Sci. Technol., 352–356 (2016)

Pandey, V.C., Abhilash, P.C., Singh, N.: The Indian perspective of utilizing fly ash in phytoremediation, phytomanagement and biomass production. J. Environ. Manage. **90**(10), 2943–2958 (2009)

Research Committee of Japanese Institute of Landscape Architecture: Ground maintenance manual in landscape planting. J. Jpn. Inst. Landscape Archit. **63**(3), 224–241 (2000). (in Japanese)

Sheoran, V., Sheoran, A.S., Poonia, P.: Soil reclamation of abandoned mine land by revegetation: a review. Int. J. Soil Sed. Water **3**(2), 1–21 (2010)

Singh, L.P., Siddiqui, Z.A.: Effects of fly ash and Helminthosporium oryzae on growth and yield of three cultivars of rice. Bioresour. Technol. **86**(1), 73–78 (2003)

Slavich, P.G., Petterson, G.H.: Estimating the electrical-conductivity of saturated paste extracts from 1-5 soil, water suspensions and texture. Aust. J. Soil Res. **31**(1), 73–81 (1993)

Srivastava, A., Chhonkar, P.K.: Amelioration of coal mine spoils through fly ash application as liming material. J. Sci. Ind. Res. **59**(4), 309–313 (2000)

Takeuchi, K., Shimano, K.: Vegetation succession at the abandoned Ogushi sulfur mine, central Japan. Landscape Ecol. Eng. **5**(1), 33–44 (2009)

Ueno, K.: A mechanism of soil acidification in acid sulfate soils. Ann. Rep. Res. Inst. Biol. Funct. **4**, 25–33 (2004). (in Japanese)

An Investigation of the Geotechnical Properties of Coal Combustion By-products from Matimba Power Station in Lephalale, South Africa

L. Magunde, F. Sengani and T. Zvarivadza

1 Introduction

Coal fly ash is a by-product of pulverised coal in an electrical generating station. Thomas (2007) pointed out that coal fly considered to be an unburned residue, which was carried away within the burning zone/area the boiler by the flue gases and eventually collected by one of the separators either mechanical or electrostatic. However, the quantity of coal ash produced depends upon the quality of coal and the method of burning the coal. Most of the fly ash is commonly used in a dry state. Most of the researchers such as Thomas (2007) and Dilip et al. (2012) indicated that fly ash disposal found to be on the wet state, which made the mode of disposal easy. Based on Thomas (2007) and Dilip et al. (2012) studies, fly ash found to be successfully used mostly in constructions projects such as road construction and also used as a supplementary cementations material for the purpose of producing cement concrete.

Bottom ash is also one of the by-products, which is generated during the process of burning coal in different thermal power plant (Dilip et al. 2012). As compared to fly ash, bottom ash has proven to be of coarse grain to fly ash, with an angular shape and mostly porous (Vijaya Sekhar Reddy et al. 2016). Furthermore, Vijaya Sekhar Reddy et al. (2016) pointed out that the bottom ash has high shear and with less compressibility. However, the proper engineering of bottom ash allows it to be used in engineering applications such as dam construction and other civil engineering-related aspects.

Eskom generates almost 95% of electricity used in South Africa. The Matimba power station is a plant that is operating under Eskom in South Africa.

L. Magunde · F. Sengani (✉)
Department of Mining and Environmental Geology, School of Environmental Sciences, University of Venda, Thohoyandou, South Africa
e-mail: fhatugeorge@gmail.com

F. Sengani · T. Zvarivadza
School of Mining Engineering, The University of the Wiwatersrand, Johannesburg, South Africa

E. Widzyk-Capehart et al. (eds.), *Proceedings of the 18th Symposium on Environmental Issues and Waste Management in Energy and Mineral Production*,
https://doi.org/10.1007/978-3-319-99903-6_12

The production of electricity at Matimba power station has been increasing over the years. Matimba power station recently produced large amount of electricity, which in turn leads to the disposal of lot of fly ash. The major problems faced by coal-based thermal power plant are handling and disposal of the coal fly (Eskom Integrated Report 2015).

There are several environmental crises associated with both fly and bottom ash, and these crises affect the life of the community members and animals around the vicinity of the thermal power stations.

This study aims to highlight some of the cardinal points that can be used to resolve the environmental crisis raised by disposal of both bottom and fly ash. The main objective of the study was achieved through detail analysis of the geotechnical properties of coal fly ash and bottom ash at Matimba power station, with purpose of utilising the ashes for construction use.

1.1 Problem Statement

In the production of electricity through the combustion of coal, Matimba power station is producing huge amount of fly ash and bottom ash. The major problems faced at the Matimba power station is the disposal and the handling of the coal combustion by-product which are the fly ash and the bottom ash whereby safe disposal and handling of large quantities of fly ash and bottom ash have not only become tedious but also expensive. To resolve the problem, strategic decision on the utilisation of coal ash has to be developed.

1.2 Objectives of the Study

The main objective of the research was to investigate the geotechnical properties of fly and bottom ash to establish their potential utilisation as a construction material. The specific objectives were to determine the grain size distribution of the fly ash, to conduct geotechnical analysis associated with soil strength and also evaluate its usability within the construction industry.

2 Research Approach

2.1 Material Sampling

Sampling was conducted of fly ash, bottom ash and the mixture of bottom and fly ash for laboratory tests. These samples were investigated to determine grain size distribution and Atterberg limits. The main purpose was to identify different types of ashes and the analysis of their geotechnical properties.

The ash samples were attained by initially augering to the depth of about 50 cm from the surface to remove the disturbed and mixed ashes. After reaching the desired depth, the auger was slowly removed from the hole and the samples were collected immediately after the removal of the auger from the hole. Samples from different locations were collected randomly. The sampling plastic bags were labelled with names, sampling date and location. The samples collected from the fly ash pipe were labelled FLY 01 and FLY 02, samples from the bottom ash chamber were labelled as BA 01 and BA 02, samples from the dump site were labelled as AD 01, AD 02 and AD 03, and, lastly, the plastering sand that was collected from the construction site was labelled as PS01.

Soil samples collected during site investigation were wrapped in plastic bags to preserve and protect them from any possible mixing or moisture changes. The samples were eventually sent to the laboratory for index tests.

2.2 Laboratory Work

The laboratory work was conducted towards an understanding of material behaviour, to determine the nature of the ashes and how they perform under the imposed conditions such as the presence of water, temperature and pressure. Analysis of site samples with respect to how conditions would change over the time provided an understanding of the existing and future conditions, as well as the suitability of material for construction purposes.

2.3 Sample Preparation

Samples that were collected from the field were initially weighed using scale balance approximately 500 g from each bag of ash samples. The samples were then taken to the oven for drying in which 110 °C temperature was used to completely remove the moisture for 24 h. The bag of samples was left for a while to cool out immediately when removed from the oven. The dried samples were then weighed and the weight was recorded for each sample. The main objective of weighing sample before drying and weighing them after drying was to determine the total moisture content of each measured sample.

2.4 Sieve Analysis

The main purpose of carrying out particle size distribution was to classify the sample materials into different categories as well as to determine their texture. This was because one of the major factors that affect the behaviour of the soil mass is the size of the grains. The apparatus used was a stack of sieves with the pan at the bottom and the cover on the upper sieve, Sartorius balance with an accuracy of 0.01 g, mechanical sieve shaker, plate and the drying oven.

To determine the grain size distribution, the samples were placed in a Vacutec oven overnight. The weight of each empty clean sieve was recorded in a notebook. The sieves were then set in a descending order starting with a 4-mm sieve on top to 0.075 mm on the bottom and a pan as the last plate at the bottom. The dried material from each bag was weighed into 500 g each. The 500 g from each sample bag was sieved through a stack of sieves with each successive screen in the stack from top to bottom having a smaller opening to capture progressively smaller particles. The materials were mechanically shaken using the sieving machine for an hour at an amplitude of 40 m. The amount retained on each sieve was collected and weighed to determine the amount of material passing each sieve size as a percentage of the total sample being sieved.

3 Results and Discussion

This chapter presents the data which were collected from the study area to achieve the objectives of the research. The data interpretation and analysis are also presented in this chapter. The results obtained from the laboratory work were used to design graphs and tables.

3.1 Water Content Test Results

Table 1 shows the percentage moisture content of different fly ashes and bottom ashes. Sample FLY 01 and FLY 02 contain a small amount of water as shown from the results in comparison with other sample ashes. This led to the conclusion that coal fly ash cannot hold much water. The results show that pure fly ashes contain on average 1% of water.

Bottom ashes, on the other hand, give different results compared to the fly ashes. In Table 1, BA 01 and BA 02 show a higher percentage of water content. The average water content for the bottom ashes shows a percentage above 40% water availability in the material. Coarse-grained materials tend to have high void spaces compared to fine grained which can contain lot of water. Therefore, it can be concluded that the bottom ashes are composed of coarse-grained materials and the

Table 1 Total moisture content

Sample no.	Mass of wet sample	Mass of dry sample	Mass of water	Moisture content (%)
FLY 01	500.01	498.41	1.6	0.32
FLY 02	503.74	503.12	0.62	0.12
BA 01	505.43	331.52	173.91	52.46
BA 02	507.06	341.76	165.3	48.37
AD 01	504.77	440.60	64.17	14.56
AD 02	513.79	445.73	68.06	15.27
AD 03	503.44	463.32	40.12	8.66
PS 01	566.26	480.95	85.31	17.74
PS 02	529.02	400.29	106.73	26.66

particle size of the sample can resemble the individual grain size of the sample based on the physical observations.

The material that is composed of the mixture of bottom ash and fly ash AD 01, AD 02 and AD 03 was collected from the damp ash. The results in Table 1 show the average percentage of the water content of the sample ranging from 8% and above. The average water content of the mixture depends from the ratio of the fly ash and bottom ash in the sample. The samples collected show that there was a higher percentage of fly ash as compared to bottom ashes.

A reasonable mixture of bottom ashes and fly ashes can make an excellent engineering property that can be used for construction and building. The mixture of ashes with a reasonable water content can provide material with good strength and stability based on the previous studies associated with performance of construction material.

3.2 *Particle Size Distribution Results*

The percentage passing of the materials on each sieve shows a gradual decrease in the overall material with decrease in the sieve sizes. The results represented in Table 2 show percentage of the samples analysed during this research. The results show that the coal fly ashes have 100% passing from the sieve size 4 to 1 mm as the material descends in the successive sieves. The 100% passing rate in these sizes proves that the materials are dominated with fine-grained materials.

The percentage passing of the bottom ashes shows gradual decrease from the upper sieve to the pan as the successive sieve size descends. The results show that the material is composed of varying grain sizes. The gradual decrease in material shows that the material is a mixture of fine-grained materials and coarsed material. The material can be concluded is composed of silt- to sand-sized materials, and this conclusion was based on grain size analysis.

Table 2 Percentage passing through each sieve

Sieve size (mm)	Percentage passing (%)							
	FLY 01	FLY 02	BA 01	BA 02	AD 01	AD 02	AD 03	PS
4	100	100	98	95	100	100	100	100
2	100	100	88	85	99	99	99	99
1	100	100	80	79	98	99	99	98
M	99	99	70	69	98	97	98	84
0.25	96	97	53	53	94	94	94	50
0.125	82	82	31	50	78	77	78	18
0.075	66	65	19	19	60	58	62	9

The mixture of fly and bottom ashes, which was collected in the damp site, shows slight decrease in percentage passing as the sieve size descends. The results from the table show that the amount of the fly ashes is greater than the amount of bottom ashes mixed. The material is dominated with fine-grained material of other coarse ones.

The mixture of materials to a reasonable extent can have advantages in terms of strength and stability in civil engineering purposes. The mixed material tends to have better sorting and decreased void spaces, which in turns increase the strength and suitability of the material for building purposes. The gradation curve of the coal fly ash, as shown in Fig. 1, shows that the material is mostly dominated by clay materials with very fine sizes. The percentage passing of the material tends to be higher from the upper sieve to the pan. The percentage passing of the material ranges ranges from 65 to 100% (Fig. 1).

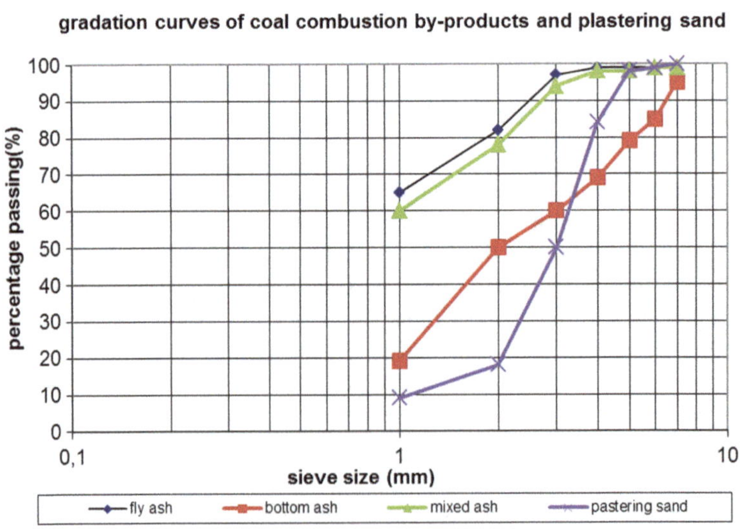

Fig. 1 Gradation curves of coal combustion by-products and plastering sand

The results of percentage passing of bottom ash are shown in the gradation curve (Fig. 1). The results show that the material is composed of silt- to sand-sized materials as the percentage passing is starting from 19 to 96%. The results of percentage passing of mixed ashes are shown in the gradation curve in Fig. 1, where the material passing is starting from above 50 to 100% from the descending sieves. The material is mostly dominated by fine clay material and a small proportion of silt-sized material. The results of plastering sand show that the material is mostly composed of silt- to fine-sand material as the gradation curve shows the size of material ranging from 9 to 100%.

3.3 Atterberg Limits

Pandian (2004) says that Atterberg limits are known to be the basic measures of water content within fine-grained soil; however, Pandian (2004) includes the following aspects: shrinkage limit, plastic limit and liquid limit.

3.3.1 Liquid Limit

The liquid limits of the coal fly ash, bottom ash and mixture of the two are presented in Table 1. The average liquid limit of coal fly ash was calculated to be 29%, which shows that coal fly ash does not contains high volume of water. The liquid limit of fly ash also shows that the material is mainly composed of fine-grained materials. The average liquid limit for bottom ash was calculated to be 71%, which means that the bottom ashes are mostly dominated by coarse-grained or sand-sized materials that contain significant amount of water. The mixture of the bottom and fly ash was calculated to be 34%. This depends on the proportion of each material added in the mixture. The average liquid limit of the mixture shows that there is higher proportion of coal fly ash compared to bottom ash. The mixture also shows that the material is mostly composed of more fine materials compared to bottom ash. The standard plastering sand shows the lowest liquid limit with an average of 15%.

3.3.2 Plastic Limits

Table 3 contains the results of the analysis of plastic limits. The average plastic limit of the coal fly ash was calculated to be 7%, which shows that the fly ash is slightly plastic (Table 3). The bottom ash has an average of 37% (Table 3). The results show that the material is highly plastic, and it is mostly dominated by silt-sized materials. The results of mixture of bottom and fly ash were determined to be 10%, which shows that there is a more fly ash than bottom ash. The material can be grouped as slightly plastic due to the proportion of fly ash in the mixture. The plastering sand is non-plastic where the plastic limit was 4%.

Table 3 Atterberg results

Samples	Average liquid limit (%)	Plastic limit	Plasticity index	Description
FLY 01 and FLY 02	29	7	22	Medium plastic
BA 01 and BA 02	71	37	34	High plastic
AD 01, AD 02 and AD 03	34	10	23	Medium plastic
PS01 and PS02	15	4	11	Slightly plastic

3.3.3 Plasticity Index

The highest plasticity index was of bottom ash which was calculated to be 34%. Bottom ash exhibits the high plasticity compared to other materials. The plasticity index of fly ash was calculated to be 22%, and the plasticity index of the mixed ashes was calculated to be 23% as shown in Table 3. The fly ash shows that the material is slightly plastic.

The plasticity index charts of fly ash, bottom ash and mixed ashes are shown in Fig. 2. Figure 2a shows the plasticity index of fly ash in which material is slightly plastic with clay nature. Figure 2b shows the plasticity of mixed material with the same plastic nature. Figure 2c shows the plasticity of bottom ash with the medium plastic nature. Figure 2d shows that bottom ash has organic silt clay material with low to medium plastic nature. The plastering sand shows that the material has low liquid limit and low plastic limit as shown in the chart above with low clay material as indicated by the A-line.

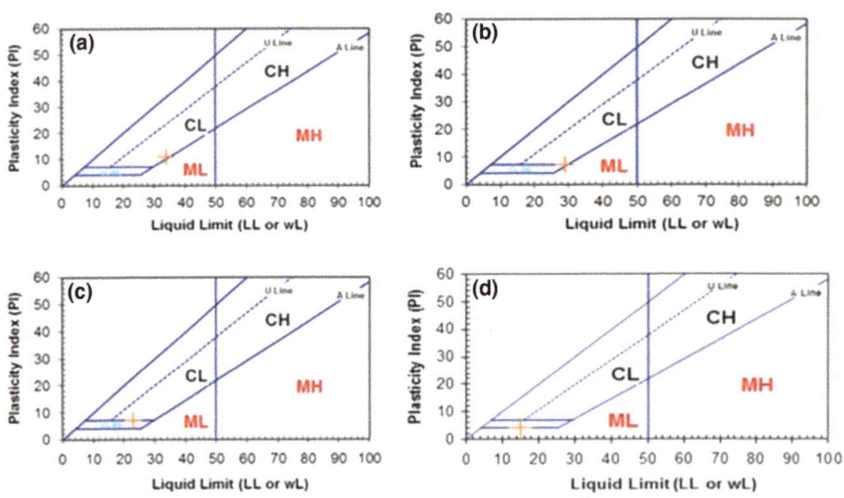

Fig. 2 Plasticity index of **a** fly ash, **b** mixed material with the same plastic nature, **c** bottom ash with the medium plastic nature and **d** bottom ash with organic silt clay material with low to medium plastic nature

Table 4 Summary of sieve analysis data and classification

Sample number	Gravel (%)	Sand (%)	Silt/clay (%)	AASHTO CLASS
FLY 01	0	34	66	A-7-6
FLY 02	0	35	65	A-7-6
BA 01	0	81	19	A-7-5
BA 02	0	81	19	A-7-5
AD 01	0	40	60	A-7-6
AD 02	0	42	58	A-7-6
AD 03	0	38	62	A-7-6
PS	0	91	9	A-7-6

3.3.4 AASHTO

The material samples were comprised of sand and clay or silt particles according to AASHTO classification of the grain sizes. The highest amount of particles in fly ash samples in silt clay is 66%, in bottom ash is 19% while in mixed ash was calculated to 60%. According to this classification system, the fly ash can be classified as clay material while bottom ash can be classified as silt to clay material. The plastering sand has a minimum of 9% clay content and 91% sandy material, as shown in Table 4.

4 Conclusions

The purpose of this research was to determine the geotechnical properties of coal fly ash and bottom ash to determine whether the material will be suitable to be used as building material. Coal fly ash at Matimba power station was determined to be clay material while bottom ash was established to be silt to clay material.

Laboratory tests conducted on material samples showed that the coal fly ash has very low moisture content, while bottom ash has high water content. Liquid limit of fly ash is 29%, bottom ash with 71% and the mixed ashes with 34%. Plastic limit of coal fly ash was calculated to be 7%, bottom ash with 34% and the mixed ashes with 10%. Low moisture content of coal fly ash renders high shear strength while bottom ash shows moderate shear strength.

Coal fly ash can be a resource rather than being an environmental problem due to its suitability for use as building material. The material has good geotechnical engineering properties, and it can be used as building material for construction purposes, specifically as plaster replacement as well as an additive to plastering material.

References

Dilip, K., Neetesh, K., Ashish, G.: Geotechnical properties of fly ash and bottom ash mixtures in different proportions. Int. J. Sci. Res. (IJSR), 1487–1488 (2012)

Eskom Integrated Report: The impact of coal ashes, South Africa, Limpopo, pp.80 (2015)

Pandian, N.S.: Fly ash characterization with reference to geotechnical applications. J. Indian Inst. Sci. **84**, 189–216 (2004)

Thomas, M.: Optimising the Use of Fly Ash in Concrete, p. 2. Portland Cement Associate, Skokie, IL (2007)

Vijaya Sekhar Reddy, M., Ashalatha, K., Sasi, K.: Strength properties of ternary blended concrete by alccofine bottom ash replacement in cement and blast furnace slag in fine aggregate. Int. J. Eng. Res. Technol. **2**, 112 (2016)

Environmental Management and Metal Recovery: Re-processing of Mining Waste at Montevecchio Site (SW Sardinia)

P. P. Manca, G. Massacci and C. Mercante

1 Introduction

The affordability of mineral raw materials is crucial for the sustainable development. In EU, however, mining activities can take place under very strict environmental regulations and constraints. Old mining and processing waste represent, in fact, a major environmental risk; thus, an action of combining environmental management with metal recovery from processing waste is becoming more and more attractive (Bellenfant et al. 2013). Tailing or dump re-processing should obtain a final tail with residual concentrations of heavy metals lower than the legal limit values by: (1) producing a final froth that meets the metallurgy constraints and (2) can be considered a marketable concentrate. These two objectives make the application of the process very different from the usual industrial beneficiation processes, for which the scavenger section is aimed at obtaining the maximum recovery in the final float, while the tail grades are simply a result. The present study case is aimed at obtaining a residual concentration that should be below the environmental threshold, while the scavenger froth grade is a consequent result. Moreover, industrial plants have the scavenger float re-circulated back into the process (to maximize recovery), while, in this case, it is a final product non-marketable and in need of an inertization process.

This paper discusses the case of the waste disposed in the Montevecchio–Ingurtosu mining district (SW Sardinia) where the main minerals extracted were sphalerite and galena in siliceous gangue. The re-treatment of these tailings raises some difficulties, depending on: (i) the fine particle size and the low content of the already floating material; (ii) the alteration processes due to the geochemical

P. P. Manca (✉) · G. Massacci · C. Mercante
DICAAR—Department of Civil and Environmental Engineering and Architecture, University of Cagliari, Cagliari, Italy
e-mail: ppmanca@unica.it

E. Widzyk-Capehart et al. (eds.), *Proceedings of the 18th Symposium on Environmental Issues and Waste Management in Energy and Mineral Production*,
https://doi.org/10.1007/978-3-319-99903-6_13

environment, through the high specific surface and time elapsed after deposition (between 50 and 80 years); (iii) the lead and zinc oxidized forms. These aspects will be discussed below along with the experimental results of batch laboratory tests.

2 Tailing Re-treatment

2.1 Froth Flotation Applicability

Froth flotation is a highly versatile method for separating physically particles based on differences in the ability of air bubbles to selectively adhere to specific mineral surfaces in mineral/water slurry (Kawatra 2011). The flotation system includes many interrelated components as shown in Fig. 1.

Collectors generally employed in flotation are surfactants that form, for instance, electrostatic bonds with the solids (Kursun and Ulusoy 2012). Difficulty arises when Pb or Zn minerals are separated from an ore having a low grade or a complex composition (complex sulfide ores) or when the mineral surface properties (oxidized Pb and Zn ores) make the response to flotation extremely poor. The design of a flotation plant, therefore, requires bench scale batch and locked cycle tests, followed by pilot plant trials (Dunne et al. 2010).

2.2 Effect of Surface Liberation and Particle Size

The extraction of ore minerals from fine-grained flotation tailings is a commercially interesting but technologically challenging endeavor, which needs to be supported by a full technical and economic feasibility study.

Jameson (2012) discussed how recovery of mineral particles by flotation is strongly a function of the particle size. As the size of floatable particles increases,

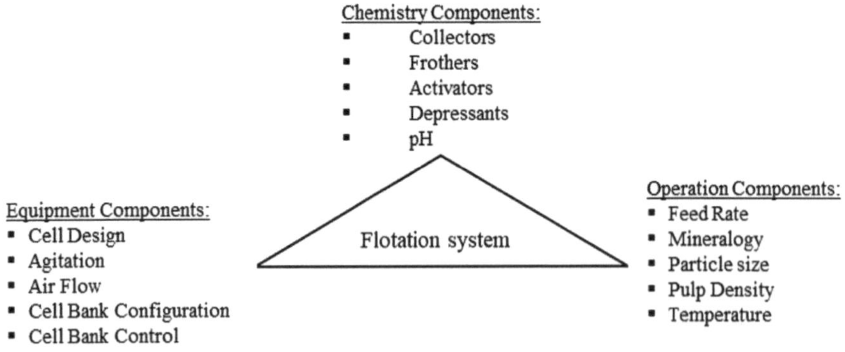

Fig. 1 Interrelated components of a flotation system [Klimpel 1995 cited in (Kawatra 2011)]

the recovery increases too, until it reaches a maximum, before decreasing monotonically.

Büttner et al. (2018) discussed the case study of a historic tailing storage facility containing on average 0.2 wt% of Sn as cassiterite. The viability of the flotation process was attributed to three material parameters, namely grade, liberation, and particle size of the mineral. These parameters were quantified for a set of ten exploration drill cores by chemical assay and mineral liberation analysis. Results illustrated the importance of combining chemical grade data with quantitative mineralogical and microfabric information to be able to assess objectively the residual value contained in the industrial tailings or any other residue to be re-processed.

2.3 Complex Pb–Zn Sulfide Ores

Complex sulfide ores are generally defined as those ores for which it is difficult to recover one or more selective products of acceptable quality and economic value having minimal losses and at reasonable costs. Complex sulfide ores are fine-grained, intimate associations of chalcopyrite, sphalerite, and galena, disseminated in dominant pyrite, and containing valuable amounts of minor elements (Marabini and Barbaro 1990).

A number of studies have been carried out on complex Pb and Zn sulfide ores. It is worth citing, as a recent example, the article of Kursun and Ulusoy (2012) illustrating the case of a selective flotation circuit that is treating (−74 μm) Pb–Zn–Cu complex ore. The ore feed to the plant consists of sulfide minerals, namely galena (5.2%), sphalerite (5.1%), chalcopyrite (0.9%), and pyrite (7.8%). The main gangue mineral is quartz. Using three stages of cleaning and three stages of scavenging, a zinc concentrate having 58.81% Zn was produced with a recovery of 74.21%. The integration of the column flotation circuit to current conventional flotation circuits of the plant will lead to an improvement in final zinc concentrate grade by a minimum of 6 and 13% points of recovery and reduction of three cleaning stages at the same time.

2.4 Oxidized Zn and Pb Minerals

There is a difference between sulfurized and oxidized mineral flotation. It is quite easy to recover Zn from sphalerite and Pb from galena even using xanthate collectors, while it is more difficult to recover Zn from smithsonite and Pb from cerussite. This topic has been discussed extensively in the relevant literature.

The flotation of Zn and Pb oxidized minerals is difficult because there are no known direct-acting collectors capable of producing single-metal concentrates. In the case of oxidized lead and zinc minerals, their surface is not easily rendered hydrophobic by the collectors generally used, to achieve efficient flotation. Furthermore, the solubility of these oxide minerals is high. Consequently, the collector also interacts with the metal cations in the solution, increasing the quantity of reagent required for flotation. It is therefore common practice to sulfurize such minerals before flotation to prepare their surface to receive xanthates, which are the collectors generally used for concentrating sulfides (Barbaro 2000). Normally, the collectors used react when the ore underwent preliminary sulfurization surface. This is an extremely delicate and critical phase. In fact, sulfurization requires a careful dosage; otherwise, the mineral surface becomes inert.

Mehdilo et al. (2012) investigated the flotation of smithsonite from a tailing derived from the cerussite flotation circuit using Armac C and Armac T collectors. The results indicated that the dosage of sodium sulfide, collector type, and desliming are the main affecting factors on the Zn grade and recovery of smithsonite flotation concentrate. Without consumption of any gangue depressants, an optimal concentrate was obtained with 94.4% recovery and 42.3% Zn grade. In such a condition, the collectors' consumption is decreased significantly. Thus, desliming before flotation is essential for a successful recovery of smithsonite.

Ejtemaei et al. (2014) reviewed an interesting analysis of zinc oxide mineral beneficiation using flotation method. Dosages of sulfidizing agent and collector, desliming before flotation, and the use of sodium hydroxide instead of sodium carbonate to act as pH regulator are essential to the effective recovery of smithsonite in cationic flotation. Despite the use of sodium silicate and SH as depressants in oleic acid flotation, flotation was not selective. Flotation using mixed collectors (Armac C+ Potassium Amyl xanthate) showed promising results. The ratio of the mixed collectors and the sequence of addition of mixed collectors were important criteria in obtaining true mixed collector flotation.

2.5 *Historical Case Studies*

Two of the most recent examples of tailing re-processing are described.

Güven et al. (2010) refer to some lead–zinc ore tailings containing 3.12% Zn, 3.43% Pb, 0.71 g/t Au, and 74 g/t Ag. Sulfide flotation was effective to obtaining bulk concentrates with high zinc content. Oxide flotation was effective for concentrates having a high lead content. Moreover, it was observed that increasing stages of rougher flotation followed by cleaning stages produced an increase in the concentrates' metal content. After sulfide flotation, a bulk lead–zinc concentrate with 43.02% Zn and 12.23% Pb was produced: The highest lead content that could be obtained by oxide flotation was equal to 22.7%.

Evdokimov and Evdokimov (2014) studied how the conversion of sulfides into marketable selective concentrates was carried out in two stages. The sulfide product is extracted from tailings first and then processed jointly with a current ore material accepted technology, or separately, by jet flotation. Sulfides are extracted from tailings using a channel-type hydroseparator.

3 The Montevecchio Mining District

3.1 Mining History

The Montevecchio–Ingurtosu–Gennamari mine basin is part of central west Sardinia (Italy). The ore deposit, developed for a length of over 10 km, consists of hydrothermal veins filling a fracture field occurring both in arenaceous–phyllitic–quartzitic meta-sediments of the post-Gothlandian series and in the Hercynian granitic batholiths (Salvadori 1973). The main metalliferous minerals are sphalerite and galena, with subordinate pyrite, chalcopyrite, covellite, arsenopyrite, and tetrahedrite. The gangue is mainly quartz, siderite, ankerite, and sporadic baryte and calcite.

The Montevecchio area was divided into three sectors: (i) Levante (East), including the Sciria, Piccalinna, and S. Antonio mines; (ii) Ponente (West), including the Sanna, Telle, and Casargiu mines; (iii) Ingurtosu, including the Ingurtosu, Gennamari, S'Acqua Bona, and Perda S'Oliu mines. The richest parts of the complex were the vein belt N60°E (Piccalinna, Sanna, Telle, Brassey), and the S. Antonio vein (with an outcrop stretching 2 km in length, and about 8 m thick). In both cases, the exploitation was 600 m deep. A vertical zonation was identified in the vein system: Galena prevailed in the upper parts, while in the intermediate zones a progressive enrichment in sphalerite occurred. An intense oxidation took place for a length of about 4 km and a depth of about 100 m, forming iron hydroxides and cerussite. In the Ingurtosu area, three main veins are encountered from west to east: the Ingurtosu vein, rich in galena; the Cervo vein (galena and sphalerite); the Brassey vein (sphalerite) (Dessau 1935). These outcrops were exploited since the Phoenician–Punic times; during the Roman occupation of the Island, silver was extracted from the argentiferous galena (De Caro et al. 2013), continuing with alternate fortunes since the Middle Ages. The mine industrial exploitation, however, dates back to the second half of the nineteenth century (Biddau et al. 2001): The first exploitation permission was received in 1848. Since then, various beneficiation plants operated in the area, concentrating the minerals by gravimetric separation, until in the 1920s the flotation process was introduced, implementing the building of new plants and the modernizing the old ones (Caboi et al. 1993).

In the period 1948–1973, from the minerals extracted in the district about 1.6 Mt of Pb and 1.1 Mt of Zn were produced, with Ag amount of 600–700 g per Pb ton, along with Bi, Sb, Cu, Cd e Ge (Salvadori 1973). One-third of the mine mineral production was extracted from S. Antonio vein. The mine activities were stopped for good in 1991.

3.2 Waste Dumps and Tailings in Montevecchio

In the Montevecchio–Ingurtosu mining area, about 150 dumps (130 of which in the West area), allowing for more than 2 Mm^3, and a big tailing dam (5 Mm^3) are found, while tailings are disposed or spread in a very large area (more than 7 Mm^3). Their total amount corresponds to more than 20 Mt. The Pb and Zn residual concentrations are both variable from 0.3 to 2.0% (Table 1).

4 Materials and Methods

4.1 Materials

The tailing mineralogical XRD analysis indicates that muscovite, quartz, and sphalerite were present. Five mixtures were considered, having a different Pb and Zn grade (from very poor tailings to a sample of sphalerite collected from the same dumps). The multi-element chemical analysis results of the feed tailings in the flotation tests are shown in Table 2. The reagents used in the flotation tests are shown in Table 3.

4.2 Methods

The flotation tests were performed in a 2.5-dm^3 Denver cell using 700 g of solid feed. The flotation flow sheet consisted of three parts (see Fig. 2): rougher, cleaner, and scavenger. Three final products were obtained: final froth (the commercial concentrate), Final tails (the reusable waste), and scavenger froth (the non-commercial concentrate). The third product should be taken into consideration too: In fact, it is not appropriate to recycle the scavenger float since it would increase the grade of the tail, while the goal is opposite. Indeed, in the scavenger section, the higher the Zn grade of the froth, the lower the Zn grade of the tails.

Samples were ground and then sieved to collect the −250 +5 μm fraction for a series of five groups of flotation tests carried out according to the flow sheet in Fig. 2. In the rougher section, $CuSO_4$ was added and the pulp was conditioned for 5 min at pH equal to 6.5. The slurry was conditioned with collectors (conditioning time 10 min). Finally, Dowfroth was used as a frother (100 g/Mg). The froth collection was carried out for 10 min. The final Froth, final tails, and scavenger froth were dried, weighed, and analyzed using a spectrometer ICP/OES (PerkinElmer 7000 Optima).

The final froth was obtained after two cleaner flotations. Before arriving to the scavenger section, the material was slimed at a high-pressure hydrocyclone (0.5 MPa, and d_{50} = 5 μm). The different grade of the two products should satisfy

Table 1 Mining and mineralogical wastes in Montevecchio district

Site	Dumps			Disposed tailings			Tailing dam			Widespread tailings		
	Volume	Average grade		Volume	Average grade		Volume	Average grade		Volume	Average grade	
	m^3	Pb%	Zn%	m^3	Pb%	Zn%	m^3	Pb%	Zn%	m^3	Pb%	Zn%
ML	61,000	1.12	0.73	700,000	2.06	2.00	5000,000	0.28	0.67	6800,000	0.31	0.25
MPI	2200,000	0.81	0.76	–			–			600,000	0.54	0.94

ML Montevecchio Levante; *MPI* Montevecchio Ponente–Ingurtosu

Table 2 Multi-element chemical analysis of the feed tailings in the flotation tests (metal grades in mg/kg)

Test N.	As	Cd	Cr	Pb	Zn	Cu
Flot_01	158	240	50	1769	29,207	206
Flot_02	164	259	50	1762	31,364	209
Flot_03	123	135	53	1806	17,339	190
Flot_04	120	126	53	1809	16,290	189
Flot_05	104	78	54	1826	10,837	181

Table 3 Reagents used in the flotation tests

Chemical	Concentration (g/t)	Plant section	Role
Na isopropyl xanthate	100	Rougher and cleaner	Sulfide collector
Cu SO_4	400		Sphalerite activator
Dowfroth 1015	100		Frother
Na_2 CO_3	400	Scavenger	pH controller
Na oleate	300–400		Oxide collector

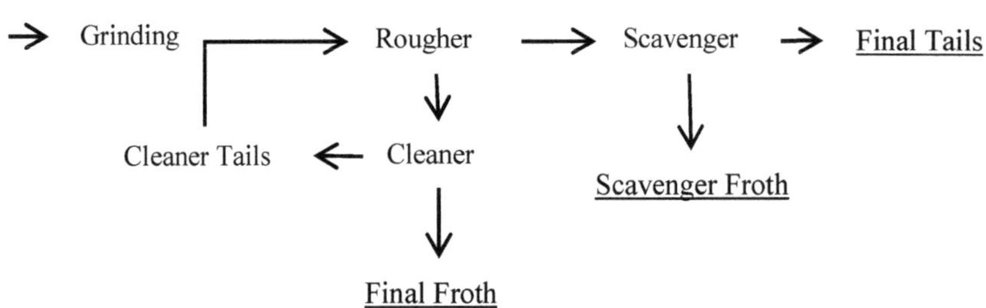

Fig. 2 Flowsheet of the flotation tests

the following conditions: The float should have Zn > 40–50% (according to the optimal metallurgical process conditions) and the final tails Zn < 1500 mg/kg (according to Italian Environmental Regulations for Industrial Sites). The scavenger froth was considered as a waste to be inertized at a minimum cost.

5 Results and Discussion

The flotation results of the five test groups are shown in Table 4 and in the histograms in Figs. 3, 4 and 5. The metal recovery in the final froth has always been between 95 and 70%, according to the modality of execution of the cleaner

Table 4 Final froth and final tails: multi-element chemical analysis results

	As	Cd	Cr	Pb	Zn	Cu
Concentration grade mg/kg						
Flot_01	366	3164	27	5208	502,003	2535
Flot_02	312	2425	56	4894	552,273	2521
Flot_03	894	99	43	11,109	506,512	88
Flot_04	711	3675	33	8900	477,833	3967
Flot_05	673	454	42	6449	446,623	466
Tailing grade mg/kg						
Flot_01	30	7	44	586	1019	72
Flot_02	28	5	31	834	1227	52
Flot_03	19	4	30	554	915	63
Flot_04	40	7	34	581	1028	68
Flot_05	40	8	35	660	1009	70

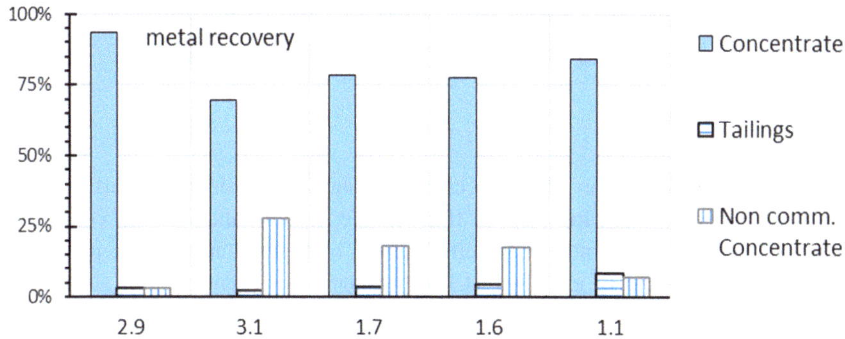

Fig. 3 Flotation results: metal recovery versus feed grade

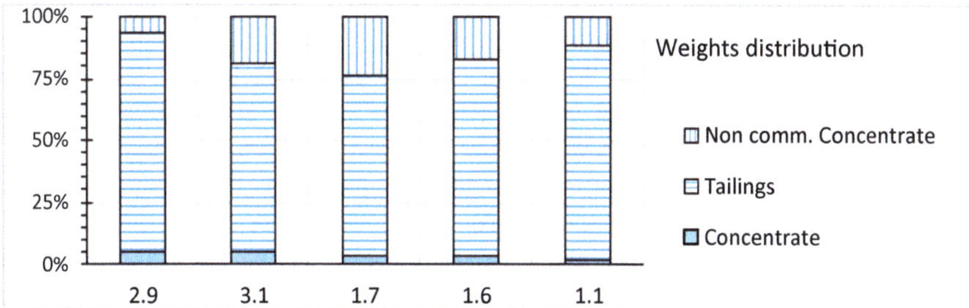

Fig. 4 Flotation results: metal recovery versus feed grade

flotation, but not for the grade of Zn in the feed. The same happened for the weight distribution. Therefore, the best flotation groups were the first and last ones, for which a multi-stage cleaner was used. As far as the final tail product is concerned,

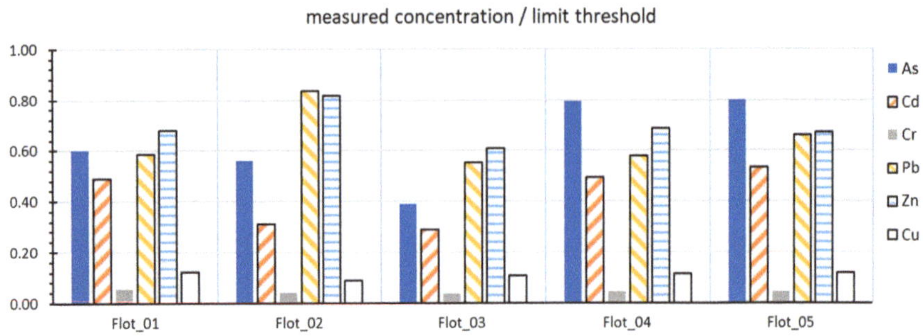

Fig. 5 Flotation results: concentration/limit threshold

compliance with the legal limits is highlighted in the histograms in Fig. 5, showing the values of the ratios between the measured concentrations and the corresponding thresholds (as required by the Italian Regulations) of the six elements analyzed. Since the values are all less than 1, the threshold value is always respected.

6 Conclusions

The mining wastes in the Montevecchio area show a considerable variability in terms of particles size, time elapsed from the deposition, and residual concentrations of lead and zinc minerals. In some dumps, untreated materials are present as they were not thought of any interest from an economic point of view sixty years ago.

The results obtained on the mixed materials are particularly encouraging in terms of both environmental management (low grade of heavy metals in the final tail) and metal recovery (market value of the concentrate). Concentrates with a grade of zinc respecting the requirements of metallurgy were obtained: They can be considered as marketable products with a commercial value (2017). The final tail material can be used for other purposes, such as for filling and landscape modeling: As an example, in the western part of Montevecchio (Ponente) surface excavations of about 1.5 Mm^3 that could be filled in with the re-processing tail are present.

However, the samples considered do not represent the totality of the Montevecchio wastes for both the relative high value of the Pb and Zn concentrations considered (between 1 and 3%) and the high recoveries obtained in the flotation tests (between 70 and 95%). On the contrary, tailing dam materials have grade lower than 1% (for both Pb and Zn) and a relevant component of oxidized minerals difficult to be floated. The assessment of the whole waste on the site will require further investigation. In the future, it is expected that it will be possible to distinguish between waste that can be re-processed and those which should be placed in a storage landfill.

Acknowledgements This research was carried out in the framework of projects conducted by Centre of Excellence on Environmental Sustainability (CESA), Sardinia, Italy.

References

Barbaro, M.: Lead and zinc ores—flotation, pp. 15–34. Academic Press, Rome (2000)

Bellenfant, G., Guezennec, A.G., Bodénan, F., D'Hugues, P., Cassard, D.: Re-processing of mining waste: combining environmental management and metal recovery? In: Fourie, A.B., Tibbett, M. (eds.) Mine closure 2013, pp. 571–582. Cornwall, United Kingdom (2013)

Biddau, R., Da Pelo, S., Dadea, C.: The abandoned mining area of Montevecchio-Ingurtosu. Rend. Sem. Fac. Sci. Univ. Cagliari Suppl. **71**(2), 109–123 (2001)

Büttner, P., Osbahra, I., Zimmermann, R., Leißnerc, T., Satgea, L., Gutzmer, J.: Recovery potential of flotation tailings assessed by spatial modelling of automated mineralogy data. Miner. Eng. **116**, 143–151 (2018)

Caboi, R., Cidu, R., Cristini, A., Fanfani, L., Massoli-Novelli, R., Zuddas, P.: The abandoned Pb–Zn mine of Ingurtosu, Sardinia (Italy). Eng. Geol. **34**, 211–218 (1993)

De Caro, T., Riccucci, C., Parisi, E.I., Faraldi, F., Caschera, D.: Ancient silver extraction in the Montevecchio mine basin (Sardinia, Italy): micro-chemical study of pyrometallurgical materials. Appl. Phys. A Materi. Sci. Process. **113**, 945–957 (2013)

Dentoni, V., Grosso, B., Manca, P.P., Massacci, G.: Old mine dumps recovery: an environmental and techno-economical challenge. In: Hu, Z. (ed.): Land Reclamation in Ecological Fragile Areas. Proceedings of the 2nd International Symposium on Land Reclamation and Ecological Restoration (LRER 2017), 20–23 Oct 2017, Beijing, PR China. CRC Press/Balkema, Leiden, pp. 453–460 (2017)

Dessau, G.: Appunti sui giacimenti minerari di Gennamari-Ingurtosu. Boll. Soc. Geol. Ital. **54**(02), 229–240 (1935) (in Italian)

Dunne, R.C., Lane, G.S., Richmond, G.D., Dioses, J.: Flotation data for the design of process plants part 1—testing and design procedures. Mineral Processing and Extractive Metallurgy **119**(4), 199–204 (2010)

Ejtemaei, M., Gharabaghi, M., Irannajad, M.: A review of zinc oxide mineral beneficiation using flotation method. Adv. Coll. Interface. Sci. **206**, 68–78 (2014)

Evdokimov, S.I., Evdokimov, V.S.: Metal recovery from old tailings. J. Min. Sci. **50**(4), 800–808 (2014)

Güven, O., Burat, F., Bulut, G., Önal, G. Evaluation of lead zinc ore tailings by flotation. In: Gülsoy, Ö.Y., Ergün, L.Ş., Can, N.M., Çelik İ.B. (eds.) Proceedings of the 12th International Mineral Processing Symposium (IMPS). Cappadocia-Nevşehir, Turkey, Oct 2010, pp. 6–8 (2010). Available at: https://www.researchgate.net/publication/256794407_Evaluation_of_lead_zinc_ore_tailings_by_flotation. Accessed 5 Jan 2018

Jameson, G.J.: The effect of surface liberation and particle size on flotation rate constants. Miner. Eng. **36–38**, 132–137 (2012)

Kawatra, S.K.: Fundamental principles of froth flotation. In: Darling, P. (ed.): SME Mining Engineering Handbook, vol. 2, 3rd edn. Littleton, CO, pp 1517–1531 (2011)

Kursun, H., Ulusoy, U.: Zinc recovery from lead–zinc–copper complex ores by using column flotation. Miner. Process. Extr. Metall. Rev. **33**(5), 327–338 (2012)

Marabini, A., Barbaro, M.: Chelating agents for flotation of sulfide minerals. In: Gray, P.M.J., Bowyer, G.J., Castle, J.F., Vaughan, D.J., Warner, N.A. (eds.) Sulphide Deposits—Their Origin and Processing, pp. 103–117. The Institutions of Mining and Metallurgy, London (1990)

Mehdilo, A., Zarei, H., Irannajad, M., Arjmandfar, H.: Flotation of zinc oxide ores by cationic and mixed collectors. Miner. Eng. **36**, 331–334 (2012)

Salvadori, I., Zuffardi, P.: Guida per l'escursione a Montevecchio e all'Arcuentu. Itinerari Geol. Miner. Giacim. Sardegna I, 29–46 (1973) (in Italian)

Part IV
Mineral Processing and Tailings Treatment, Recycle, Disposal

Universal Flotation Reagent Produced from Plant Waste

S. Yefremova, L. Bunchuk, Yu. Sukharnikov, E. Li, A. Niyazov, S. Shalgimbayev, Yu. Hain and A. Zharmenov

1 Introduction

Mining and metallurgy is a main industry of economy of Kazakhstan. It is well known that nowadays quality of raw materials has worsened. There is a task to provide a complex processing of mineral resources to increase quantity and quality of the final product. Flotation is one of the methods of low-quality mineral ore processing (Aleksandrova et al. 2015; Koca et al. 2017). In this connection, the requirements to flotation methods of mineral raw materials enrichment are redoubled. One of the ways to solve this task is to create new effective flotation reagents (Han et al. 2017; Ai et al. 2017). However, any production of a flotation reagent is not in Kazakhstan. Therefore, it was important to produce it using local cheap raw materials.

The unique raw material was chosen. It was rice husk because it was renewable and common in the world raw resources.

The complex rice husk processing technology was reported (Efremova 2012). It was shown that the liquid organic product (LOP) produced from rice husk and used as a blowing agent instead of the known reagent T-92 allows increasing the lead recovery by 1.1% and improving the quality of lead concentrate by at least 0.5%

S. Yefremova (✉) · L. Bunchuk · Yu.Sukharnikov · A. Zharmenov
National Center on Complex Processing of Mineral Raw Materials,
Almaty, Kazakhstan
e-mail: s_yefremova@cmrp.kz; secretar_rgp@mail.ru

E. Li · A. Niyazov · S. Shalgimbayev
«Kazmekhanobr» State Scientific Production Association of Industrial Ecology,
Almaty, Kazakhstan

Yu.Hain
ICMD International Corporation of Metal and Alloy Development Holding GmbH,
Hannover, Germany

E. Widzyk-Capehart et al. (eds.), *Proceedings of the 18th Symposium on Environmental Issues and Waste Management in Energy and Mineral Production*,
https://doi.org/10.1007/978-3-319-99903-6_14

Table 2 Results of Akzhal ore pilot testing in experimental mode with using of composition TM 067:LOP = 1:1

Products	Yield (%)	Content (%)		Recovery (%)	
		Pb	Zn	Pb	Zn
Lead concentrate	0.95	57.92	12.62	84.84	2.18
Zinc concentrate	8.10	0.54	61.15	6.74	90.20
Dump tail	90.95	0.06	0.46	8.42	7.62
Ore	100.0	0.65	5.49	100.0	100.0

Table 3 Results of Akzhal ore pilot testing in base mode with using of TM 067 Danafloat

Products (%)	Yield (%)	Content (%)		Recovery (%)	
		Pb	Zn	Pb	Zn
Lead concentrate	0.955	57.72	13.90	84.81	2.41
Zinc concentrate	8.14	0.55	60.77	6.88	89.98
Dump tail	90.905	0.06	0.46	8.31	7.61
Ore	100.0	0.65	5.50	100.0	100.0

The recovery of lead in the concentrate of the same name amounted to 84.84%. When TM 067 Danafloat was used, the content of lead and zinc in the lead concentrate was 57.72 and 13.90%, respectively, with the concentrate yield being 0.955% (Table 3). In the case of base mode, the recovery of lead into the concentrate was slightly lower (84.81%), while the loss of zinc in the lead concentrate was, on the contrary, higher (2.41% vs. 2.18%) compared to the experimental mode.

The results of zinc flotation are not in favour of the base mode as well. The values of such indications as metal content and recovery of zinc into the concentrate obtained in the base mode (zinc content—60.77%, lead—0.55%, recovery of zinc into the concentrate—89.98%, Table 3) are inferior to those obtained in experimental mode (content of zinc—61.15%, lead—0.54%, recovery of zinc into the concentrate—90.20%, Table 2). However, the yield of the concentrate in the base mode (8.14%) was slightly higher than in the experimental mode (8.10%).

4 Conclusions

Due to the presence of compounds containing hydrophobic and hydrophilic groups of atoms, the liquid organic product obtained from the rice husk shows the properties of a collecting agent and as previously stated of a blowing agent. Flotation of lead–zinc ores using LOP ensures high process parameters of enrichment corresponding to the use of commercial flotation reagent TM 067 Danafloat as well as better ones. Based on the obtained results, it can be concluded that replacing 50% of consumption of TM 067 Danafloat with a liquid organic product from rice husk during the process of flotation of Akzhal lead–zinc ore provides lead concentrate quality enhancement by 0.20% while the content of zinc in it is reduced by 1.28% and zinc concentrate—by 0.38%, with an increase in zinc recovery by 0.22%. Taking into account that the cost of LOP is $3.7 USD/kg lower than the cost of TM 067 Danafloat, using LOP in composition with TM 067 Danafloat 1:1 will allow saving $0.1 on each ton of ore. With the capacity of 1200 thousand tons of ore per year, the factory will save substantial $120 thousand. So, along with being economically attractive the liquid organic product from rice husk is a competitive universal flotation reagent for enrichment of polymetallic lead–zinc ores.

Acknowledgements We acknowledge the Ministry of Education and Science of the Republic of Kazakhstan (Project No. 2252/GF4) for the financial support of the current work.

References

Ai, G.H., Huang, W.F., Yang, X.L., Li, X.B.: Effect of collector and depressant on monomineralic surfaces in fine wolframite flotation system. Sep. Purif. Technol. **176**, 59–65 (2017)

Aleksandrova, T.N., Nikolaeva, N., Romashev, A.: An experimental and theoretical approach to the assessment of the specific surface of apatite-nepheline ore in the process of grinding. In: Science and Technologies in Geology, Exploration and Mining (SGEM 2015). Proceedings of the 15th International Multidisciplinary Scientific Geoconference, vol. I, pp. 577–583. Technology Ltd., Albena (2015)

Efremova, S.V.: Rice hull as a renewable raw material and its processing routes. Russ. J. Gen. Chem. **5**(82), 999–1005 (2012)

Han, H.S., Hu, Y.H., Sun, W., Li, X.D., Cao, C.G., Liu, R.Q., Yue, T., Meng, X.S., et. al.: Fatty acid flotation versus BHA flotation of tungsten minerals and their performance in flotation practice. Int. J. Min. Proc. **159** 22–29 (2017)

Koca, S., Aksoy, D.O., Cabuk, A., Celik, P.A., Sagol, E., Toptas, Y., Oluklulu, S., Koca, H.: Evaluation of combined lignite cleaning processes, flotation and microbial treatment, and its modelling by box Behnken methodology. Fuel **192**, 178–186 (2017)

Yefremova, S., Sukharnikov, Y. Bounchuk, L., Kablanbekov, A., Li, E., Niyazov, A., Shalgimbayev, S., Zharmenov, A.: Development of a new flotation reagent based on rice husk. In: Ghodrati, B., Kumar, U., Schunnesson, H. (eds.) Mine Planning and Equipment Selection (MPES 2017). Proceedings of the 26th International Symposium (2017)

Yefremova, S.V., Bunchuk, L.V., Li, E.M., Niyazov, A.A., Sukharnikov, Y.I.: Semi-industrial tests of flotation reagent from rice husk as collector. Complex Use Miner. Resour. **4**, 5–11 (2017)

Laboratory Studies to Examine the Effects of Adding Cement to Various Layers of a Surface Paste Tailings Storage

S. Tuylu, A. Bascetin and D. Adiguzel

1 Introduction

In recent years, there has been a significant amount of research conducted in thickened and paste tailings storages. These studies have shown that surface paste tailings storage can minimize the free water amount, which is one of the biggest issues associated with traditional tailings dams. Additionally, surface paste material can be stored in layers under atmospheric conditions and dried quickly. Thus, by keeping below the saturation value, the risk of liquefaction under the dynamic loads is reduced (ICOLD and UNEP 2001; Aswathanarayana 2003; Fourie 2003; Cooke 2008). At the same time, cement binder can be added to the surface paste material when necessary to increase durability and stability and reduce the risk of heavy metal mobilization with acid mine drainage (AMD). If cement or another binder is going to be used in the paste material mixture, the amount of binder (cement) by solid weight is desired to be $\leq 2\%$ in terms of cost and being able to store more tailings (Benzaazoua et al. 2004; Deschamps et al. 2011; Bascetin and Tuylu 2017; Bascetin et al. 2017a, b).

Studies of surface paste disposal conducted in industrial areas throughout the world have been more in the form and have not yet come into practice as an alternative method (Theriault et al. 2003). As the experiments regarding the subject are new, an efficacious model for the optimum storage design has not yet been developed. It has been observed that the potential environmental impacts in the literature have generally been analysed geochemically (AMD, etc.) in terms of the column, humidity cell and small-scale cabin tests (Verburg 2001, 2002; McGregor and Blowes 2002; Benzaazoua et al. 2004; Deschamps et al. 2007, 2008, 2011).

S. Tuylu · A. Bascetin (✉) · D. Adiguzel
Mining Engineering Department, Faculty of Engineering,
Istanbul University - Cerrahpasa, Istanbul, Turkey
e-mail: atac@istanbul.edu.tr

E. Widzyk-Capehart et al. (eds.), *Proceedings of the 18th Symposium on Environmental Issues and Waste Management in Energy and Mineral Production*,
https://doi.org/10.1007/978-3-319-99903-6_15

The main aim of this study was to determine the optimal design that allows for successive layers of storage in a laboratory environment, taking into account the physical and geochemical durability of mine process tailings. One of the other aims of the study is to enable the paste material to achieve a physical durability feature in order for it to be able to remain stable under its own weight or under static or dynamic loads according to the subsequent use of the storage area. In addition, the prevention of water and oxygen diffusion within the stored material after the storage is completed is another important issue. A storage method that complies with these conditions will be able to minimize environmental risks and be a suitable method for reclamation and reutilization conditions later on.

In this context, after the determination of the physical and chemical characteristics of Pb–Zn tailings, mixtures between 65 and 75% solid content (SC) by weight have been prepared. The process tailings material and water, which are the components of the paste material, are mixed to obtain a homogeneous mixture suitable for pipe transportation. The preparation temperature of the samples was 22 ± 3 (°C). For this mixture to be pumped for long distances, slump tests have been conducted in order to achieve a slump value of 250 mm (acceptable for surface paste disposal), and accordingly, the optimal SC value was determined. In order to fully understand the behaviour of the paste waste material during pouring, three laboratory-type experimental paste tailings deposition cabins in Fig. 1 (containers of dimensions 2 m length × 0.7 m width × 1.3 m height) have been used to allow the simulation of the paste waste material in surface storage. Paste was poured into each cabin in 11 layers of equal thickness and with 8 days between pouring each consecutive layer. Measurement sensors were placed in layers 1, 5 and 10 inside the cabin in order to instantaneously measure the oxygen consumption, matric suction and volumetric water content (VWC) values of the paste material.

In the study, the surface behaviour of paste material was simulated with all layers being uncemented in Design 1 (reference cabin), followed by some layers being cemented in Design 2 and Design 3 storage cabins, and the changes in oxygen, matric suction and volumetric water content values in the paste material

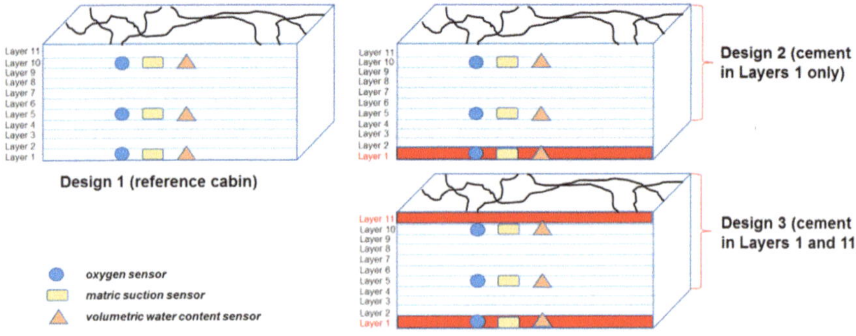

Fig. 1 Test cabin designs with different configurations

during pouring were determined. Thus, the optimal storage design for Pb–Zn tailings prepared for SPD was identified.

2 Materials and Methods

The paste material used in this study was obtained by mixing the lead–zinc process tailings with the water used in the facility. While preparing the paste, CEM I 42.5 Portland cement (PC) produced in accordance with the EN 197-1 standard (British Standards Institution 2011) was used as a binder in certain configurations in order to research the effect of the binding material on the paste. The solid–water ratio of the paste layers was determined according to a slump value of 250 mm within the scope of this study. Additionally, cement binder in the amount of $\sim 2\%$ by solid material weight was added for the cemented layers.

Storage designs were conducted initially with Design 1 (reference cabin) with all layers without binders, followed by Design 2 and Design 3 in which cement binder ($\leq \%2$ solid content weight) was used in certain layers (Fig. 1).

Design 1 consisting of the test cell filled with this test cabin is a control sample for the other test cabins where different configurations were utilized. Therefore, this cabin can also be called the "reference cabin".

Design 2 consists of only the bottom layer (first layer) with cemented paste waste material with all the subsequent layers above it being uncemented paste tailings material. By creating an alkali medium with the cement structure in the first layer, the aim was to keep the high pH value of the pore water and prevent the formation of acid or metal mobilization. At the same time, the binder material increased the stability of the paste in the face of excessive loads on the lowest layer by strengthening the bonds between grains.

In Design 3, layers 1 and 11 were poured with cemented paste material with all the layers in between being stored as uncemented paste material. Thus, it was believed that by increasing the stability of the topmost layer of the storage area, it could be beneficial for the reclamation works after the mine was closed. It was also believed that the high pH value of the topmost layer which would come in contact with free surface waters would reduce the risk of AMD.

The paste material was placed in layers in these test cabins. The layer height of the paste material poured into the cabins was determined to be 40 mm for each layer by taking into consideration the spread width of the paste and the studies conducted in the literature (Benzaazoua et al. 2004; Dechamps et al. 2008, 2011). Paste storage comprised a total of 11 layers placed within the cabin. Additional layers were placed only once the previous layer had completed its curing time and until there was no volumetric water content loss—a time period of 8 days. Data collection from the installed sensors were completed over 88 days, while these sensors measure the volumetric water content, matric suction and oxygen consumption of the layers. This approach allowed for the previous layer to achieve stability. Three different sensors were placed in layers 1, 5 and 10 of each cabin in order to determine the values of volumetric water content, oxygen consumption and matric suction. The sensors were placed as the paste material was being placed.

3 Results and Discussion

The volumetric water content for Layer 1 and the correspondingly calculated values in the solid content by weight (SC = solid content/total mass of pulp) value are given in Figs. 2, 3 and 4 for Design 1, Design 2 and Design 3, respectively.

While the SC of Layer 1 in Design 1 shown in Fig. 2 (this layer does not contain cement) was 75.7% during initial pouring, it was observed to increase to ~82% after 88 days considering there was an opportunity of drainage. The upper layers that were poured after 40 days were determined to have little effect on the bottom layer of the reference (non-binder) paste storage system and remained stable in the range of ~80–82% (Fig. 2). The volumetric water content behaviour of the first layers of Design 2 and Design 3 consisting of binded (cemented) paste material is shown in Figs. 3 and 4. It is clearly seen that the first layers of both designs depict a similar trend and reach a stable range after day 32 with each subsequent layer pouring.

As shown in Figs. 3 and 4, after the first layer is poured, an increase in volumetric water content of ~35% is observed. It was determined that Design 2 remained stable between 83 and −86% SC and Design 3 remained stable between 81 and –84% SC after the pouring of the fifth layers. These results can be expressed as the paste material being consolidated with the loads upon it and reaching hydrostatic balance after a certain period of time (Figs. 2, 3 and 4). This situation has been observed especially after half of the total layer pouring time had passed.

Immediately after the pouring of the first layers in Design 2 and Design 3, it is thought that one of the most important causes regarding the increase in VWC was

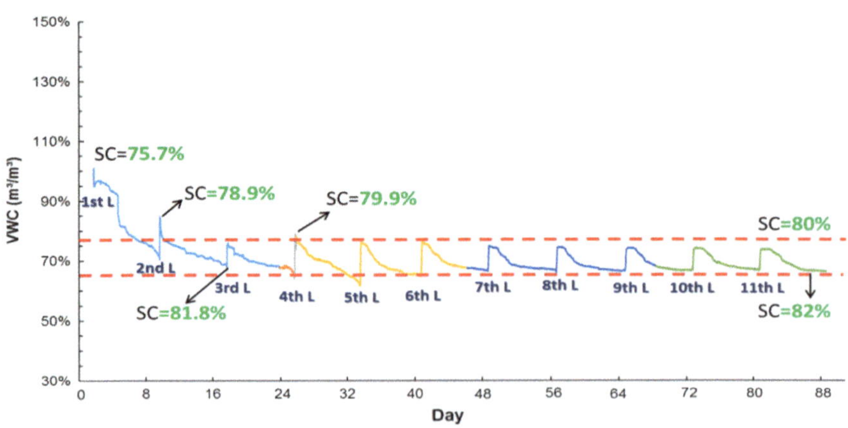

Fig. 2 VWC and SC values of Layer 1 in Design 1 (no cement)

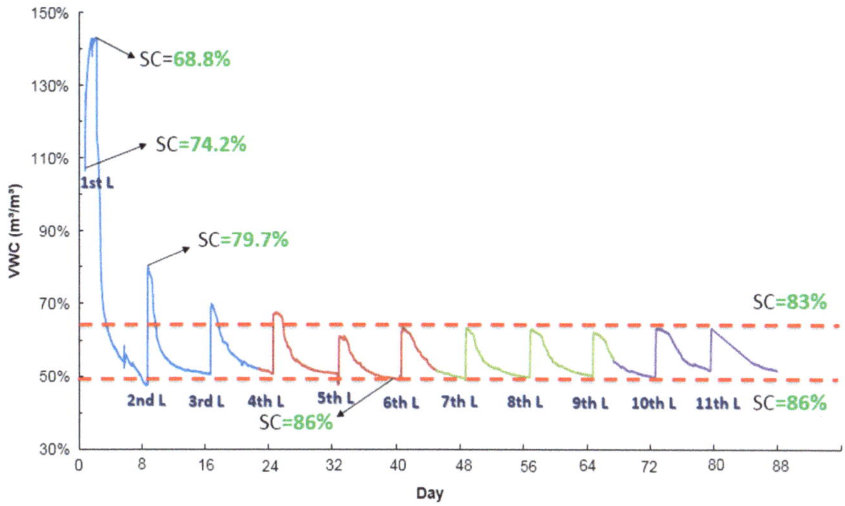

Fig. 3 VWC and SC values of Layer 1 in Design 2 (cement in Layer 1 only)

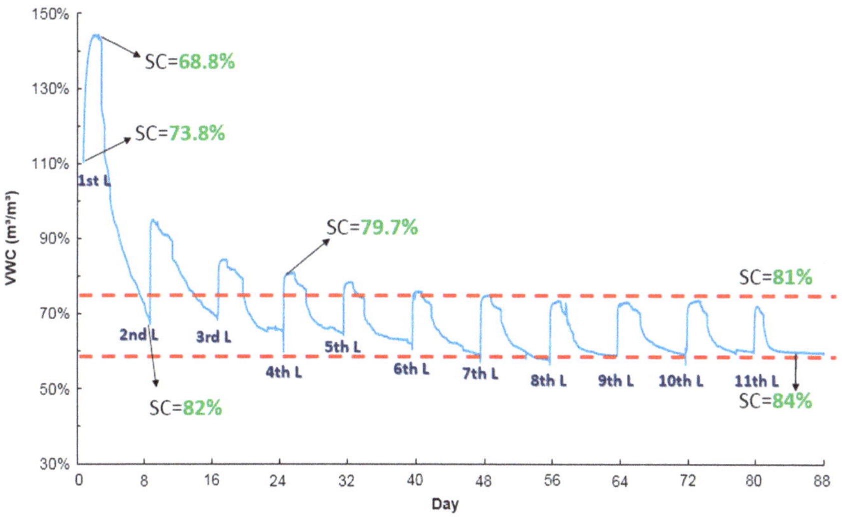

Fig. 4 VWC and SC values of Layer 1 in Design 3 (cement in layers 1 and 11)

due to the volumetric increase in C–S–H gels formed in the hydration of cement. In addition, a thaumasite salt containing 15 molecules of water may be formed as a result of the sulphates in the medium acting on the calcium silicates (C–S–H) and may cause the softening of the material by expanding. Also, the mixtures formed with Portland cement, when coming in contact with sulphated waters, tricalcium aluminate (C_3A) turns into ettringite by combining with calcium sulphate ($CaSO_4$).

This material, also called Candlot salt, may crack the material and cause it to break due to the vast volume it occupies. However, as the Design 2 and Design 3 remained stable in higher SC values when compared to Design 1, it is believed that C–S–H gels formed instead of thaumasite and ettringite salts due to the comparatively low amount of cement added and formed a solid bond between the grains. It is therefore understood that the capillary cracks are formed within its own structure thus preventing larger cracks from forming and enabling them to remain more saturated in comparison with Design 1 from the VWC values obtained from the sensor measurements of Layer 5 of each design shown in Figs. 5, 6 and 7.

As shown in Fig. 5, the fifth layer in Design 1 appears to be stable at ~87–89% SC after the initial pouring. It can be seen from Figs. 6 and 7 that the fifth layers of Design 2 and Design 3 show similar behaviours as their lowest layers are both cemented. The fact that these two designs have a SC value of about 4% less than that of Design 1 means that the fifth layer and the other underlying layers are more saturated. As can be seen from the graphs in Figs. 6 and 7, it was determined that the VWC varied over a range of about 25% throughout each pouring and drying period after the pouring of the fifth layer. This value is five times greater in comparison with Design 1. It is therefore understood that water seepage after each formation of paste layer in Design 2 and Design 3 penetrates the lower layers much more.

Within the scope of the study, it was measured that there was no significant difference in the volumetric water contents of the tenth layers in the cabin experiments of the three different designs (Figs. 8, 9 and 10). It has been determined that the PKO value gradually approaches 90% in a short time depending on the optimum water content of the tenth layers in these three cabins.

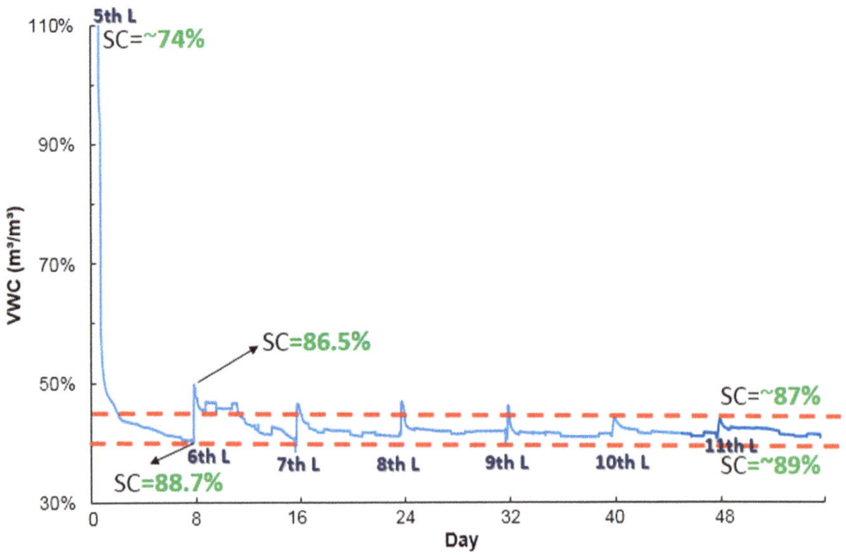

Fig. 5 VWC and SC values of Layer 5 in Design 1 (no cement)

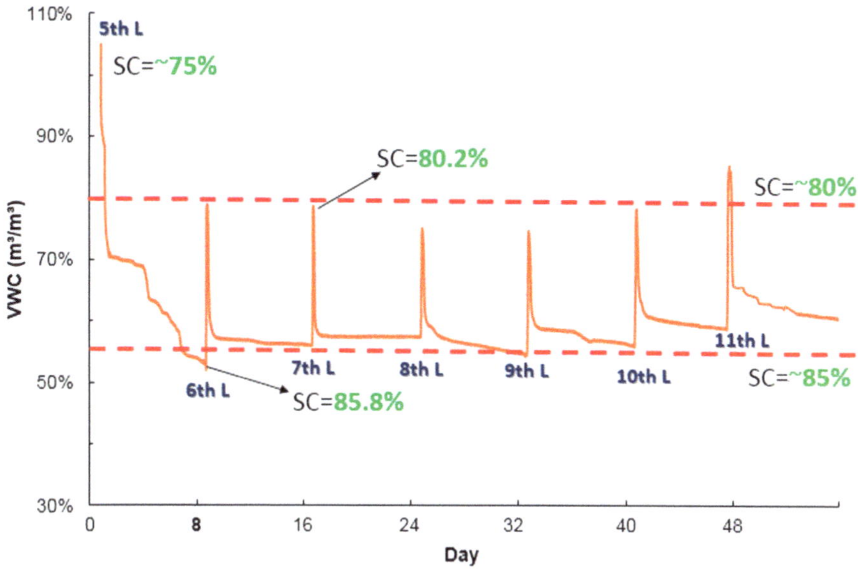

Fig. 6 VWC and SC values of Layer 5 in Design 2 (cement in Layer 1 only)

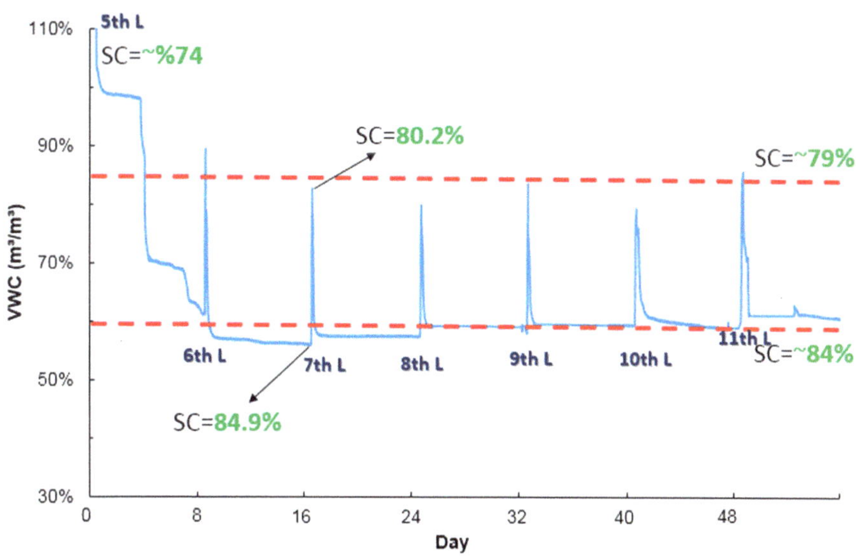

Fig. 7 VWC and SC values of Layer 5 in Design 3 (cement in layers 1 and 11)

It was determined that the SC value of the tenth layer in Design 1 remained within ~87–88% for 16 days (Fig. 8). At the same time, the SC value of the tenth layer in Design 2 was measured to change in the range of ~86–88% and the SC value of the tenth layer in Design 3 changed in the range of ~87–88% (Figs. 9 and 10). It is seen

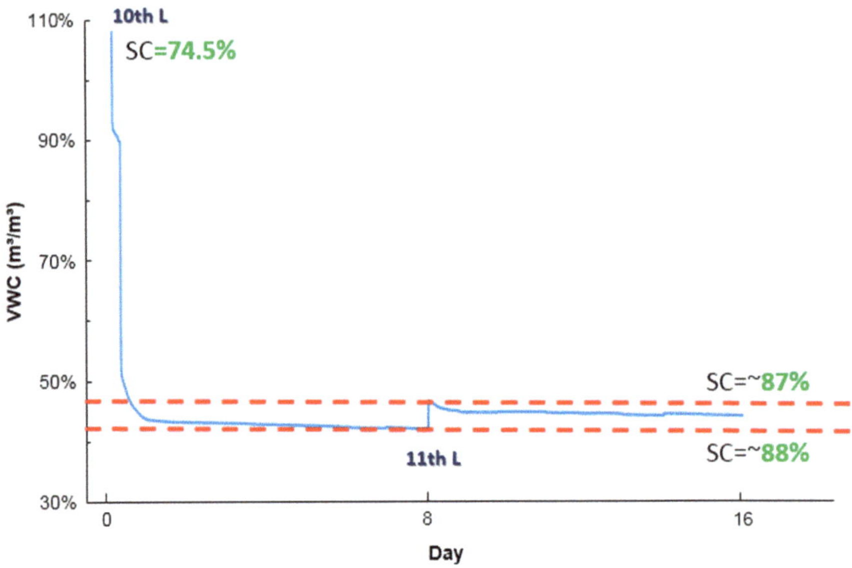

Fig. 8 VWC and SC values of Layer 10 in Design 1 (no cement)

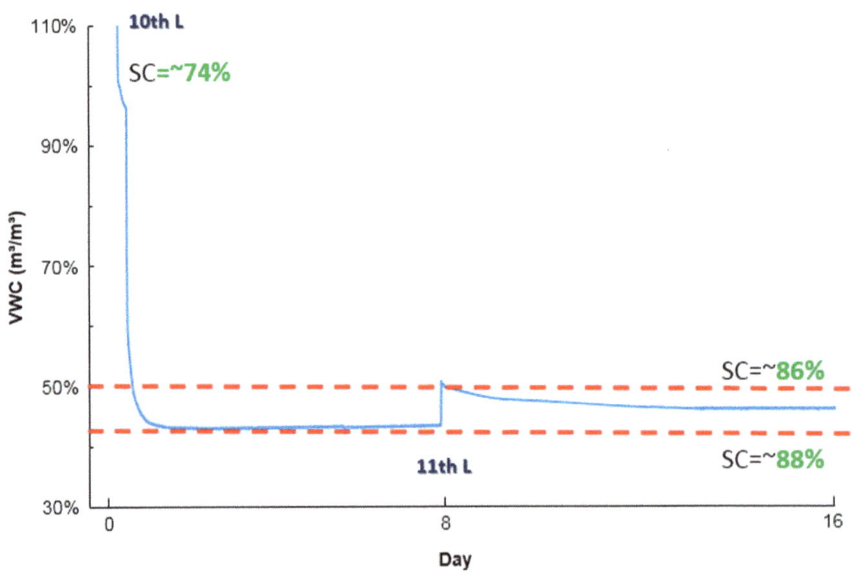

Fig. 9 VWC and SC values of Layer 10 in Design 2 (cement in Layer 1 only)

in Figs. 8 and 9 that the volumetric water contents of the tenth layers rose by 4.5% in Design 1 and 6.5% in Design 2 within about 2 h from the 11th layer being poured. However, the volumetric water content in Design 3 increased by ~5.5% within

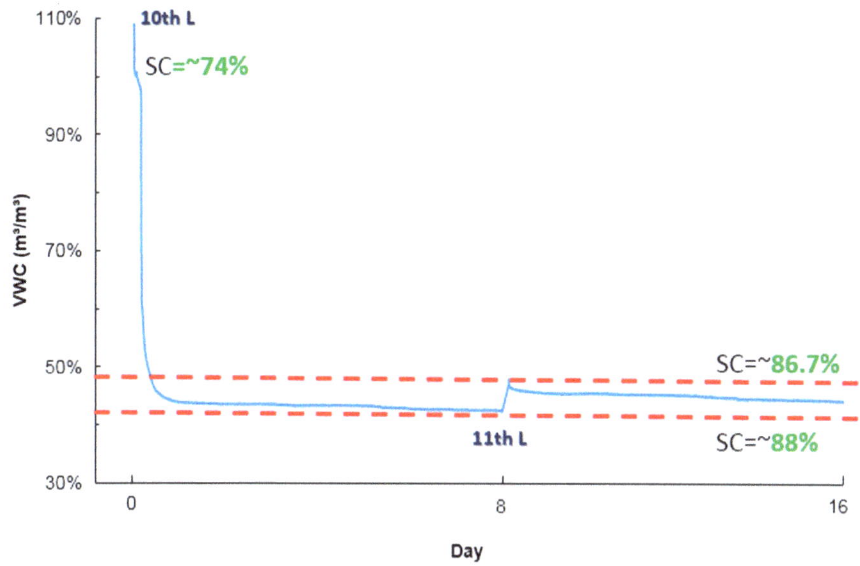

Fig. 10 VWC and SC values of Layer 10 in Design 3 (cement in layers 1 and 11)

about 9 h of the 11th layer being poured (Fig. 10). It can be stated that the passage of seepage water of the 11th layer occurs over a long time period because of the addition of cement to Layer 11 in Design 3.

When the volumetric water contents of the first, fifth and tenth layers of three different designs are examined from the geotechnical point of view, it can be seen in Figs. 2, 3 and 4 that the first layers have the greatest risk for liquefaction and shear stress when the loads on storage are rapidly consolidated and drained. The water contents of the first layers remain constant at 18% (SC ~ 82%) in Design 1 and about 14% (SC ~ 86%) in Design 2 and 3, and the water content value of %16.6 (SC ~ 83.4%) of the total volume was approached in Design 1, whereas Design 2 and Design 3 fell below this value. Also, as the number of layers increased, the effect of the matric suction between the layers of the paste material was observed and it was found that the tenth layers of three different designs stabilized at ~88% SC until the end of the casting period.

The other test results which of the matric suction pressure and oxygen values of layers belong to each design are summarized (Tuylu 2016).

The matric suction value of the first layer (−10 kPa) in Design 1 does not change during the pouring period of 11 layers. Although the volumetric water contents in the fifth and tenth layers are the same, a slight increase in matric suction is measured. It can be said that this increase is due to the fact that matric suction is directly proportional to capillary height. The matric suction pressure in the first layer in Design 2 increases from −10 to −87 kPa due to the hydration-induced matric suction requirement formed during the initial pouring because the first layer is cemented. However, with the second pouring, it was determined that the first layer remained at -10 kPa and did not change. In contrast to Design 1, the fifth layer appears to have no matric suction. Therefore, Layers 1 and 5 in Design 2 were measured to not fall below the ~50% volumetric water content in the pouring process, which is the 100% saturation limit, thus preventing the formation of suction stresses in these layers. On the other hand, the matric suction of the tenth layer started to increase 6 days after the 11th layer was poured and reached −14 kPa at the end of pouring. Here, it is thought that the fall in volumetric water content to ~46% is effective.

The matric suction values in Design 3 behave almost like the layers in Design 2. However, it is thought that the matric suction value of the tenth layer in Design 3 is higher than the tenth layers of other designs because the 11th layer of Design 3 is cemented and the capillary cracked structures that are formed after a while on the tenth layer increase the grain surface tensions after drying. Additionally, the matric suction of the tenth layer started to increase 3 days after the 11th layer was poured upon it and reached a value of -14 kPa at the end of pouring.

It is understood that the bottom layer in Design 1 comes into contact with oxygen in the air due to the cracking shortly after the initial pouring. However, it was measured that after each layer pouring followed by the pouring of the third layer, the oxygen content within the first layer decreased to 17.5–16.5% and it increased again during the 8 days drying process and remained at levels of 19.8%. The reason for this is thought to be that the seepage water remains in the layer for a while and cuts the air contact with the water-saturated paste material.

It was found that the oxygen content of the fifth layer tended to decrease until the pouring of the final layer and the lowest value was about 17%. This implies that the contact with oxygen in the air is cut off by being between the four underlying layers and the six overlying layers. In the pouring of the 10th and 11th layers, the oxygen content of the 10th layer was measured to change by very little. However, the oxygen content, which falls to ~18%, approached the oxygen value of 20.95% in the air after 16 days. This increase also occurs in a similar way in the first and fifth layers. This increase is thought to be caused by the cracks forming from the surface of the paste material to the inside due to the increase in water absorption potentials of the layers as well as the surface stresses of the particles. The oxygen content of the first layers of Design 2 and 3 shows a sudden decrease to levels of 14% after the initial pouring since they are cementitious. At the end of the pouring process, as the oxygen content values remain below 20.95% from the measurements of the first, fifth and tenth layers of these two designs, it is indicated that contact with air does not occur.

4 Conclusions

In this study, three different designs were prepared in the laboratory for surface paste disposal and the volumetric water content, matric suction and oxygen consumption values for each design were measured. In this context, it was measured that the first layers remained above the saturation limit of 53% volumetric water content in three different designs. When the fifth layers were examined, it was determined that the volumetric water content value fell below 50% in Design 1, whereas it was well above the saturation limit in Design 2 and Design 3. Also, as the number of layers increased, the effect of the matric suction between the layers of the paste material was observed and it was found that the volumetric water content of the tenth layers of three different designs was stabilized at 40–50% until the end of the pouring period. Therefore, as stated in the study by Yılmaz et al. (2014), it can be explicitly expressed with this study that the cemented layer takes on a barrier role due to its lesser and slower permeability, keeping the upper layers saturated. Since the diffusion of oxygen to the layers with a higher degree of saturation occurs at a minimum level, the AMD risk of these layers is reduced. However, this is thought to lead to stabilization problems. Theriult et al. (2003) reported in their study that the matric suction values measured near the surface of the stored paste material layer varied between 4 and 20 kPa within a few days. This corresponds to matric suction values in the layer pouring process of the tenth layer close to the surface in Design 1, which is stored without a binder. It is observed that the volumetric water content value decreases rapidly immediately after the pouring of the paste layers, remains constant for a while and then somehow slowly starts to decrease again. This basis of this change can be expressed due to the time-dependent evaporation and the diffusion of the pore waters away from the medium found between the particles of paste material that is consolidated under its own load.

References

Aswathanarayana, U.: Mineral resources management and the environment. In: Control Technologies for Minimizing the Mining Environmental Impact. Taylor & Francis (2003) (Chapter 8)

Bascetin, A., Tuylu, S.: Application of Pb–Zn tailings for surface paste disposal: geotechnical and geochemical observations. Int. J. Min. Reclam. Environ. (2017). Article in Press. https://doi.org/10.1080/17480930.2017.1282411

Bascetin, A., Tuylu, S., Adiguzel, D., Ozdemir, O.: Field properties and performance of surface paste disposal. In: Yilmaz, E., Fall, M. (eds.) Paste Tailings Management. Springer, Cham (2017a)

Bascetin, A., Tuylu, S., Ozdemir, O., Adiguzel, D., Benzaazoua M.: An investigation of crack formation in surface paste disposal method for pyritic Pb–Zn tailings. Int. J. Environ. Sci. Technol. (2017b) https://doi.org/10.1007/s13762-017-1380-5

Benzaazoua, M., Perez, P., Belem, T., Fall, M.: A laboratory study of the behaviour of surface paste disposal. In: Proceedings of the 8th International Symposium on Mining with Backfill, Beijing, China (2004), 180–192

British Standards Institution: EN 197-1: cement—part 1: Composition, specification and conformity criteria for common cements. European Standard (2011)

Cooke, R.: Pipeline design for paste and thickened tailings systems. In: Tailings and Mine Waste, pp. 95–100 (2008)

Deschamps, T., Benzaazoua, M., Bussière, B., Belem, T., Aubertin, M.: The effect of disposal configuration on the environmental behaviour of paste tailings. In: MINEFILL2007 in Montreal, Quebec, p. 2491 (2007)

Deschamps, T., Benzaazoua, M., Bussière, B., Aubertin, M., Belem, T.: Microstructural and geochemical evolution of paste tailings in surface disposal conditions. Miner. Eng. **21**, 341–353 (2008)

Deschamps, T., Benzaazoua, M., Bussière, B., Aubertin, M.: Laboratory study of surface paste disposal for sulfidic tailings: physical model testing. Miner. Eng. **24**, 794–806 (2011)

Fourie, A.B.: In search of the sustainable tailings dam: do high-density thickened tailings provide the solution. School of Civil and Environmental Engineering. University of the Witwaterstrand, South Africa, p. 12 (2003)

ICOLD and UNEP: Bulletin 121: tailings dams—risk of dangerous occurrences. In: Lessons Learnt from Practical Experiences, p. 144. Paris (2001)

McGregor, R.G., Blowes, D.W.: The physical, chemical and mineralogical properties of three cemented layers within sulfide-bearing mine tailings. J. Geochem. Explor. **76**, 195–207 (2002)

Tuylu, S.: Determination of optimum design parameters for storage in surface of mine process tailings. Ph.D. thesis. Turkey (2016)

Verburg, R.B.: Paste technology for disposal of acid-generating tailings. Min. Environ. Manage. **84** (2002)

Verburg, R.B.: Use of paste technology for tailings disposal: potential environmental benefits and requirements for geochemical characterization. In: IMWA Symposium. Belo Horizonte, Brazil (2001)

Theriault, J., Frostiak, J., Welch, D.: Surface disposal of paste tailings at the Bulyanhulu gold mine. In: Proceedings of the 2nd Mining Environment Conference. Sudbury, Ontario, pp. 1–8

Numerical Modelling of Pb-Zn Mine Tailing Dam Based on Soil Stability

A. Bascetin, S. Tuylu, D. Adiguzel, H. Eker and E. Odabas

1 Introduction

Today, there is also an increase in mining activities in order to meet the increasing human needs. As a result of mining activities, large quantities of raw materials are being produced, as well as larger quantities of tailings. Unless it is possible to utilize these tailings, they need to be stored. The tailings which are stored in bulks or in large dams can generate critical impacts on environmental and human health and safety as a result of collapse of the dams (Commission of the European Communities 2001).

The control and configuration of this kind of environmentally dangerous tailings in terms of isolation at disposal site, construction of impermeable layers, transportation from plant, stabilization, safety, their effects on water and soil quality are the main parameters that could be considered carefully.

There are some tailing disposal methods which have been applied in the mining industry for a long time. Tailing dams that are generally used in the disposal of mining tailing also bring about many problems in terms of cost, environment and human health. The number of accidents seen in such structures has started to increase along with the increase in use of this method (Cetiner et al. 2006). Although there are many reasons for accidents seen in tailing dams, the most common ones are unusual weather, seismic liquefaction, structure defects and mismanagement. More analyses should be performed in order to prevent accidents based on such reasons (Fig. 1).

A. Bascetin (✉) · S. Tuylu · D. Adiguzel · E. Odabas
Engineering Faculty, Mining Engineering Department, Istanbul University, Istanbul, Turkey
e-mail: atac@istanbul.edu.tr

H. Eker
Engineering Faculty, Mining Engineering Department, Gumushane University, Gumushane, Turkey

E. Widzyk-Capehart et al. (eds.), *Proceedings of the 18th Symposium on Environmental Issues and Waste Management in Energy and Mineral Production*,
https://doi.org/10.1007/978-3-319-99903-6_16

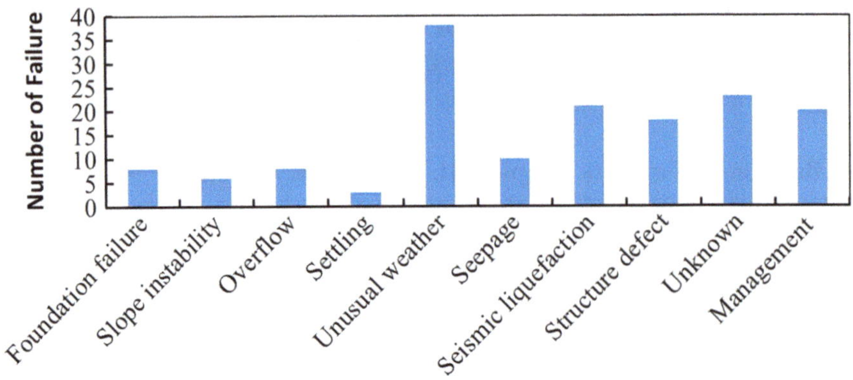

Fig. 1 Distribution of the accidents in tailing dams by their reasons (Rico et al. 2008)

As it is seen in Fig. 1, major reasons for tailing dam accidents are bad weather conditions and seismic loads. The dam must be designed in real dimensions; the necessary calculations must have been made and constructed in full capacity before starting any disposal activity. For this reasons, the complex security problems of dam construction can be technically solved. Within the scope of this study, the numerical modelling of the tailing dam of a lead–zinc mine was performed with finite element method. The most likely earthquake load (magnitude: 5.4) to be faced within the mine area was applied over the model. In addition, the effect of duration of the same seismic activity in different designs was examined on the tailing dam. Therefore, the relationship between maximum displacement and earthquake duration was revealed on different designs.

2 Materials and Methods

In this study, Pb-Zn underground mine is located in the west of Turkey; Balikesir and materials of this mine were used for numerical modelling. The Pb-Zn ore is fed into the mineral processing plant in this mine. The tailing that is produced at the end to a flotation process is directly pumped into the tailing dam without going through any other process (Bascetin et al. 2016). Furthermore, the data which were used in the numerical analysis were carried out in the previous studies, and this data were presented in Table 1 (Bascetin et al. 2015; Odx 2009; Tuylu 2016; Plaxis 1997).

The geometry and sizing of the tailings dam depend on landfill storage capacity and geographical conditions of the region. The height of the tailings dam, which was calculated as 30 m in case of the full use of the capacity, has been increased to 35 m by adding reclamation studies, a cover layer and the safety factor. As a part of the study, a database to be used as the modelling input was created with the data in Table 1 and the design parameters were calculated. Different types of embankment

Table 1 Data which use for numerical analysis

Material parameter	Unit	Embankment	Tailing	Soil	Clay	Gravel	Sand
Unit weight (Ɣunsat)	kN/m^3	18	18	20	17.28	19.165	20
Unit weight (saturated) (Ɣsat)	kN/m^3	20	21	21	18.065	19.636	21
Young's modulus (E)	kN/m^2	5.0E+04	5E+04	1.2E +06	4.3E +04	1.4E +05	6.3E +04
Poisson's ratio (v)	–	0.2	0.2	3	0.2	0.2	0.2
Cohesion (c)	kN/m^2	5	0	20	33.5	5	2
Friction angle (φ)	°	35	32	45	13	55	32.5
Dilatancy angle (ψ)	°	0	0	15	0	0	2.5

(downstream tailings dam and upstream tailings dam) in different angles of the slope were modelled in accordance with the tailings dam design planned, and the displacements that may occur in tailings dams were determined. For upstream and downstream building methods of the same capacity, designs in slope angles of 25°, 30° and 35° are created. Each of these designs is exposed to earthquake load (magnitude 5.4) with the duration of 15 s. Maximum displacement data according to different earthquake durations and the areas where those values are created are obtained as a result of the analyses.

3 Results and Discussion

In this study, the maximum displacement values were determined for the remaining six designs (the embankment of the downstream tailings dam with the slope angles of 25°, 30° and 35° along with the embankment of the upstream tailings dam with the slope angles of 25°, 30° and 35°). Accordingly, the maximum displacement values were given in Figs. 2, 3, 4 and 5.

As it can be seen in Fig. 2, in both designs made by 25° slope angle, maximum displacement values increase by increasing the duration of the earthquake. However, there have not been any significant changes in maximum displacement values for when the duration of earthquake was more than 3 s. Moreover, there is no big difference between upstream and downstream designs in terms of the effect of the earthquake.

As it is seen in Fig. 3, in both designs made by 30° slope angle, maximum displacement values increase by increasing the duration of the earthquake. On the other hand, as in the design formed by 25° slope angle, no significant changes are observed in maximum displacement values for when the duration of earthquake is more than 3 s. For when the duration of the earthquake is 3 s, the maximum displacement value for upstream design is calculated as 0.01 m and for downstream design as 0.05 m.

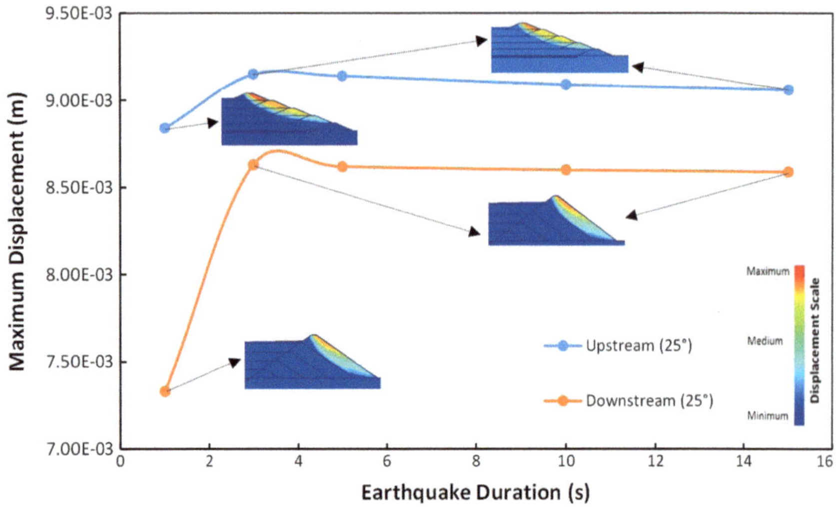

Fig. 2 The relationship between maximum displacement and earthquake duration (slope angle: 25°)

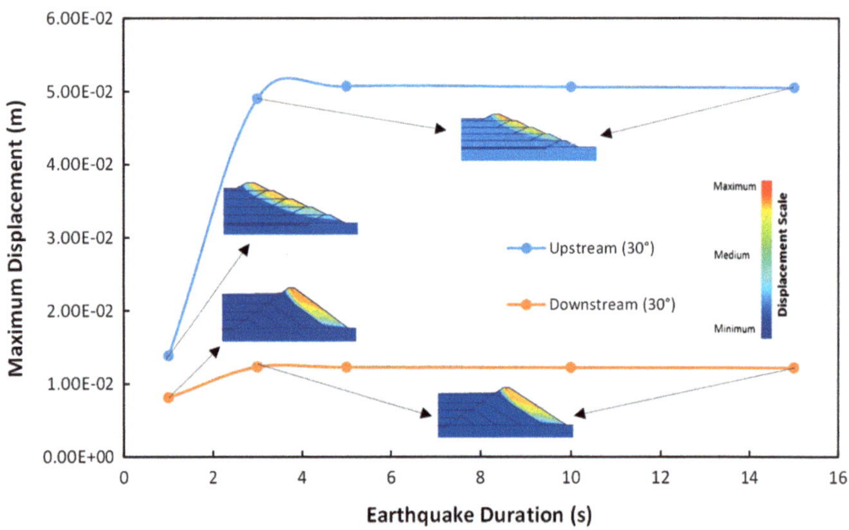

Fig. 3 Relationship between maximum displacement and earthquake duration (slope angle: 30°)

As it can be seen in Fig. 4, in both designs made by 35° slope angle, maximum displacement values increase by increasing the duration of the earthquake. However, in both designs, the maximum displacement value is largely increased with the duration of the earthquake. Accordingly, for when the duration of the

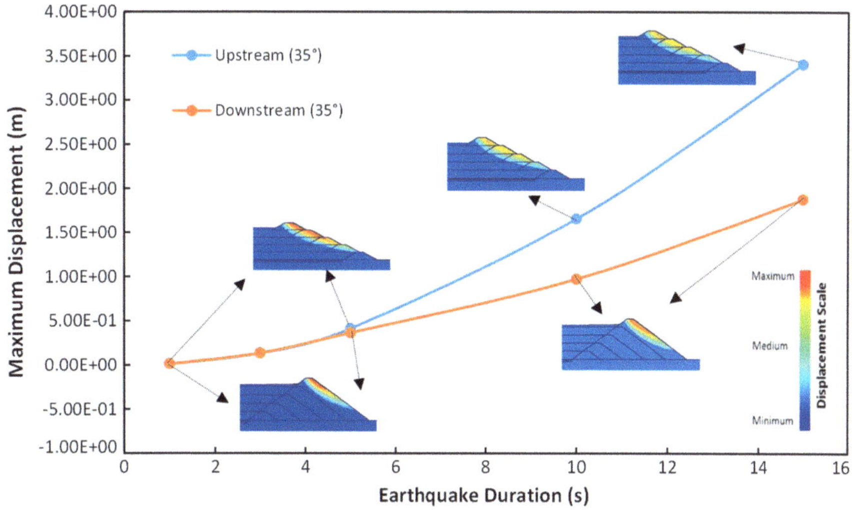

Fig. 4 Relationship between maximum displacement and earthquake duration (slope angle: 35°)

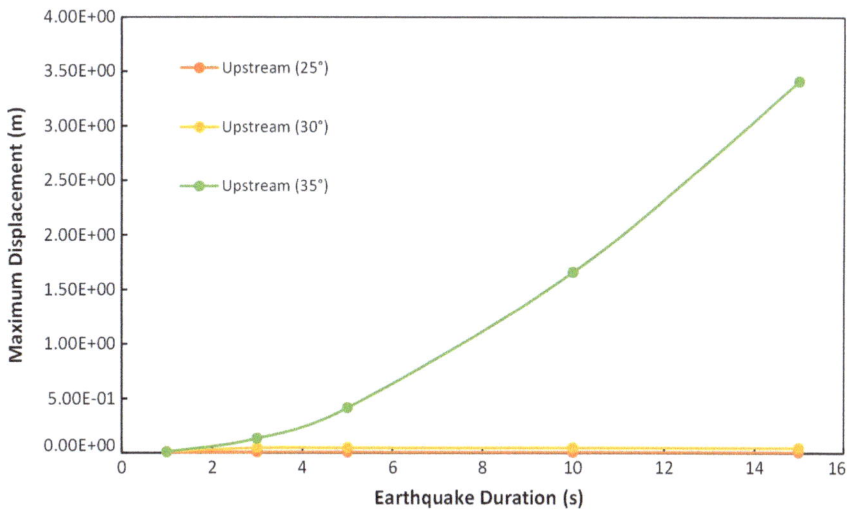

Fig. 5 Relationship between maximum displacement and earthquake duration for upstream tailings dam

earthquake is 15 s, the maximum stress value for upstream design is calculated as 3.5 m and for downstream design as 1.5 m.

As it can be seen in Fig. 5 and 6, in both designs formed by 35° slope angle, maximum displacement values have significantly increased by increasing the duration of the earthquake. This increase has been more limited for designs of 25°

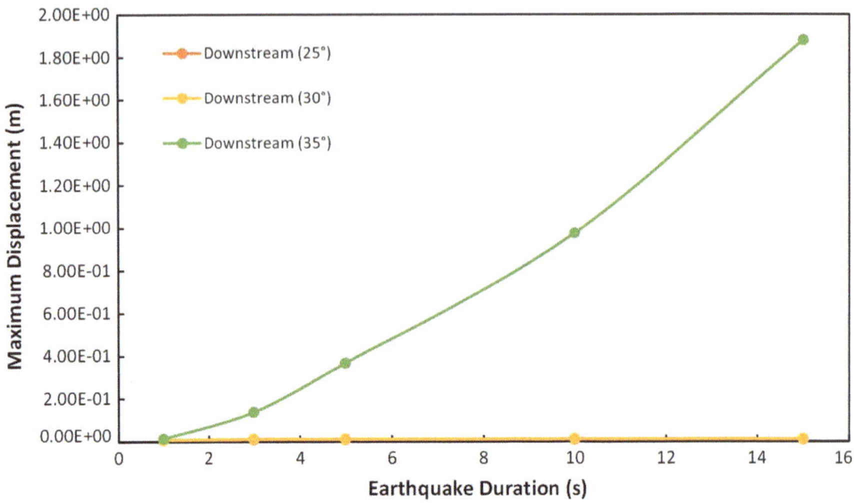

Fig. 6 Relationship between maximum displacement and earthquake duration for downstream tailings dam

and 30°. Accordingly, it can be said that the border slope angle value is 30° in terms of seismic loads for both designs.

4 Conclusions

Within the scope of this study, two different types of tailings dams were compared with the numerical modelling method under seismic loading. The displacement values which are the programme outcomes were obtained as a result of numerical modelling. Different types of embankment (downstream tailings dam and upstream tailings dam) in different angles of the slope were modelled in accordance with the tailings dam design planned. It was observed that the type of tailing dam and the duration of earthquake directly affected the maximum displacement value formed on the ground. The upstream dam is displaced 1.5–2 times more than the downstream dam at similar times. It is necessary to know the geomechanical properties of the ground on which the tailings dam are likely to be placed. It is important to design the tailings dam by considering these properties. In this study, an attempt to find an optimum tailings dam design was made by generating different scenarios under seismic loading. As a result of the data obtained, it is very important to establish the economic and security stability. Furthermore, the downstream tailings dam gives better results without any change in the slope angle in terms of safety and it is seen that the limit slope angle is 30° for the seismic loading.

References

Bascetin, A., Ozdemir, O., Tuylu, S., Adiguzel, D., Akkaya, U.G.: Storage of mining processing tailings at surface using paste technology. MT Sci. J. Undergr. Resour. (2015) pp. 33–49

Bascetin, A., Tuylu, S., Adiguzel, D., Ozdemir, O. New technologies on mine process tailing disposal. J. Geol. Resour. Eng. (2016)

Cetiner, E.G., Unver, B., Hindistan, M.A.: Mine waste related legislation: the European union and Turkey. In: TMMOB Chamber of Mining Engineers Mining Magazine, Ankara (2006)

European Comission: Management of mining, quarying and ore- processing waste in he European union. Bureau de Recherches Geologiques et Minieres BRGM (2001)

Odx: Tailing storage facility design report. Olympic Dam Expansion Project, BHP Billiton (2009)

Plaxis BV. Baldonado, M., Chang, C.-C.K., Gravano, L., Paepcke, A.: Plaxis v8.6 Software, (Build 1942), Copyright 1997–2007. The stanford digital library metadata architecture. Int. J. Digit. Libr. **1**, 108–121 (1997)

Rico, M., Benito, G., Salgueiro, A.R., Díez-Herrero, A., Pereira, H.G.: Reported tailings dam failures, a review of the European incidents in the worldwide context. J. Hazard. Mater. (2008)

Tuylu, S.: Determination of optimum design specifications for surface paste tailings disposal. Ph. D. thesis. Advisor Prof. Dr. Atac Bascetin and co-advisor, Prof. Dr. Mostafa Benzaazoua, Istanbul University Institute of Science (2016)

In Situ Evaluation of Mechanical Properties of Phosphate Tailings for Exploring Reuse Potential: Case Study of a Phosphate Mine, South Africa

T. T. Mayisa, M. E. Nengovhela, F. Amponsah-Dacosta, F. Sengani and T. Zvarivadza

1 Introduction

To obtain the metals and other minerals needed for industrial processes, fertilizers, homes, cars and other consumer products, large quantities of rock are mined, crushed, pulverized and processed to recover metal and other mineral values (USEPA 1994). Historically, most of the mining industry that extract variety of deposits in hard rocks tend to produce waste material in small particle sizes as micrometres and less during separation process; however, these particles are considered to be tailings. Although tailings are waste material, in geotechnical terms tailings also form part of soil group, such as gravel, sand, silt and clay. These different types of soil grouping help an engineer in strategic decision when rating the performance of different group of soil types (Murthy 2002).

Commonly, two distinct classification systems are used: the American Association of State Highway and Transportation Officials (AASHTO) and the Unified Soil Classification System (USCS) (Holtz 1981).

The AASHTO system of classification is based on particle-size distribution, liquid limit and plasticity index of the material. The material is classified into seven major groups, A-1 to A-7, although sometimes an organic soil is called out as A-8 (Fang and Daniels 2006). Materials classified under groups A-1, A-2 and A-3 are granular materials of which 35% or less of the particles pass through the No. 200 sieve. Materials of which more than 35% pass through the No. 200 sieve are classified under groups A-4, A-5, A-6 and A-7. These materials are mostly silt

T. T. Mayisa · M. E. Nengovhela · F. Amponsah-Dacosta · F. Sengani (✉)
Department of Mining and Environmental Geology, School of Environmental Sciences, University of Venda, Thohoyandou, South Africa
e-mail: fhatugeorge@gmail.com

F. Sengani · T. Zvarivadza
School of Mining Engineering, The University of the Witwatersrand, Johannesburg, South Africa

E. Widzyk-Capehart et al. (eds.), *Proceedings of the 18th Symposium on Environmental Issues and Waste Management in Energy and Mineral Production*,
https://doi.org/10.1007/978-3-319-99903-6_17

and clay-type materials (Murthy 2002). The members of each group have similar load-bearing values and engineering characteristics under normal traffic conditions. The best soils for road subgrades are classified as A-1, the next best A-2, etc., with the poorest soils classified as A-7. Groups A-1 to A-3 soils possess, in the densified state, an effective sand-size granular skeleton. Groups A-4 to A-7 soils possess no such bearing skeleton, and their engineering behaviour is governed by water affinity and amount. Group A-2 is subdivided into A-2–4 to A-2–7 subgroups: the last number identifying the type of minus #200 sieve fraction present (Fang and Daniels 2006). Differentiation between the qualities within a certain group is made by the group index, (IG or GI). The group index is a function of liquid limit, LL, plasticity index, PI and the per cent passing the #200 sieve, F. Then the group index can be determined by Eq. (1):

$$\mathrm{GI} = (\mathrm{F} - 35)\,[0.2 - 0.005(\mathrm{LL} - 40)] + 0.01(\mathrm{F} - 15)(\mathrm{PI} - 10). \tag{1}$$

where: GI—group index, F—% passing #200 sieve, LL—liquid limit, and PL—plasticity index (Das 2006).

The USCS system is based on the recognition of the type and predominance of the constituent material considering grain-size distribution for coarse materials and plasticity property for fine-grained materials (Holtz 1981). It divides soil into three major groups: coarse-grained soils, fine-grained soils and highly organic (peaty) soils. In the field, identification is accomplished by visual examination for the coarse-grained soils and a few simple hand tests for the fine-grained soils. In the laboratory, sieve analysis and the Atterberg limits can be used to identify the first two groups. The peaty soils are readily identified by colour, odour, spongy feel and fibrous texture (Murthy 2002).

1.1 Coarse-Grained Materials (Gravels)

The coarse-grained soils are subdivided into gravels and gravelly soils and are denoted by the symbol G, as well as sands and sandy soils denoted by symbol S. Gravel soil is considered to be the one with high percentage of course grain particle that is larger than 4.75 mm diameter, and this can be retained/contained within sieve number 4. Sands are those having the greater portion passing the #4 sieve but being retained in the #200 sieve. Both the gravels (G) and sands (S) groups are subdivided into four subgroups: GW and SW, GP and SP, GM and SM and GC and SC depending on the grain-size distribution as well as the nature and quantity of fines in the soil (Holtz 1981).

Furthermore, well-graded (W) soils have a good representation of all particle sizes whereas the poorly graded (P) soils are either uniform or gap-graded. Whether the gravel or sandy soil is well graded can be determined by plotting the grain-size distribution curve and computing the coefficient of uniformity (CU) and the

Table 1 Gravel material classification (Das 1997)

Material type	Group symbol	Group description
Gravel	GW	Well-graded gravel, gravel–sand mixture, with little or no fines
	GP	Poorly graded gravel, gravel–sand mixture, with little or no fines
	GM	Silt gravel, or gravel–sand–silt mixture
	GC	Clayey gravel or gravel–sand–clay mixture

coefficient of curvature (CC) (Murthy 2002). These coefficients are determined by Eqs. (2) and (3):

$$C_u = D_{60}/D_{10}. \tag{2}$$

$$CC = (D_{30})^2/(D_{10} \times D_{60}). \tag{3}$$

where D_{10}—grain diameter at 10% passing, D_{30}—grain diameter at 30% passing and D_{60}—grain diameter at 60% passing by mass (or weight).

The GW and SW are generally well-graded gravelly and sandy soils with less than 5% passing the #200 sieve. Holtz and Kovacs (Holtz 1981) pointed out that sand soil that contains little or no plastic fines are rated as GP and SP. Table 1 summarizes the classification of gravel soils by USCS.

1.2 Fine-Grained Materials

The fine-grained soils, having more than 50% passing the #200 sieve, are further subdivided into silts (*M*) and clays (*C*) liquid limit and their plasticity index. Organic soils are also classified as fine-grained soils even though no grain-size range is readily defined (Holtz 1981).

Fine-grained material exhibits a poor load bearing capacity. Strength changes with change in moisture condition. Laboratory classification criteria are based on

Table 2 Classification of fine-grained material (Das 2006)

LL value	Group symbol	Group description
Silt or clay with liquid limit less than 50	Ml	Inorganic silt and very fine sand, rock flour, silt or clayey fine sands or clayey silt with slight plasticity
	Cl	Inorganic clay of low to medium plasticity, gravel–clay, sand–clay, silt or clean clay
	Ol	Organic silt and organic silt clay of low plasticity
Silt or clay with liquid limit (LL) greater than 50	MH	Inorganic silt, micaceous or diatomaceous fine sandy or elastic silt

the relationship between the LL and the PI (Das 2006). Table 2 summarizes the classification of fine-grained soils.

2 Research Problem

The significant amount of mine tailings from mining operations at mine A has led to growing concerns about their ecological and environmental impacts such as the occupation of large areas of land, generation of dust, contamination of surface and groundwater. There has been a growing need to find alternative uses of tailings material to ease of the burden imposed by the tailings maintenance demands. Finding reuse opportunities of these tailings is essentially a good riddance. According to Ahmari (2012), the tailings materials are a problem in several ways: they are an economic burden, requiring significant maintenance; they pose an environmental threat through potential pollution of water resources and the land; they consume a lot of surface lands and are bad for aesthetics of the area.

3 Methodology

The field work comprised of two main activities that were conducted at the field of study: DCP testing and sample collection. Methods and techniques as well as materials used to conduct the above-mentioned activities are described in this section.

3.1 DCP Testing

In the field, the material that forms part of the dam embankment was subjected to a DCP test at various sampling points along the perimeter of the dam. This material has naturally consolidated to form a firm dam wall around the pool. The test was aimed at evaluating the bearing pressure and penetration index of the material, which is useful in the interpretation of the material's strength and load-bearing capabilities. Testing involved walking around the dam's perimeter along various test points. A standard DCP was assembled on site. The DCP consists of a hammer with a mass of 8 kg, fall distance of 575 mm per drop, as well as 60° cone at the bottom of the steel rod. A measuring rod with millimetre scale also forms part of the equipment and was used to take readings after each and every drop.

3.2 Sampling

Random sampling along the perimeter of the dam wall, at each test point, was conducted. Sampling was correlated with testing, with every sample collected at each bench from the bottom one right to the top bench. Samples were collected using a garden hand spade and were initially put in 5L bucket containers for ease of carrying from the site to the mine premises. The material was then put in plastic sample bags and was tagged accordingly. The spade was used to dig through at each sampling point to collect the fresh part of the material. Each sample was then attached with a tag, which indicated where the sample was taken and also a sample number, which provides correlation of the sample to the test. Samples were tied inside the bag, enclosing any moisture present.

3.3 Laboratory Work

The laboratory work was conducted to characterize material by grain sizes, to get an understanding of material behaviour, to determine the type of material making up the tailings and hypothesizing on how the material might behave under imposed loads using various classification systems. Such objectives were realized with the use sieve analysis technique.

3.4 Sieve Analysis

The main purpose of carrying out sieve analysis (gradation test) was to evaluate the tailings material's particle-size distribution and determine their texture. This was done because one of the major factors that affect the behaviour of material is the size of the constituent grains. The apparatus used were stack of sieves with the pan at the bottom and the lid on the upper sieve, Sertorius balance with an accuracy of 0.01 g, mechanical sieve shaker, as well as the drying oven (Vacutec oven). To determine the grain-size distribution, the samples were placed in a Vacutec oven and were dried overnight at 110 °C. The dried soil sample was sieved through a stack of sieves with each successive screen in the stack from top to bottom having a smaller (approximately half of the upper sieve) opening to capture progressively smaller particles. The amount retained on each sieve was collected and weighed to determine the amount of material passing that sieve size as a percentage of the total sample being sieved.

4 Results and Discussion

This section represents the results of the study based on the collected data from the field, as well as data from lab analysis which is meant to meet the objectives of the study. Both data analysis and interpretation are therefore covered extensively in this chapter.

4.1 Dynamic Cone Penetrometer Test Results

The results of the DCP have been applied in various ways such as soil classification, the determination of physical and mechanical soil properties, the estimation of soil bearing capacity, prediction of soil settlements and the design of shallow and deep foundations (Zein 2007). The data from DCP test yielded information about the penetration indices, whilst simultaneously providing information about bearing capacities through calculated in situ CBR.

4.2 Tailings Penetration Indices

The penetration indices from these test results were observed via graphical expressions on excel spreadsheets. Penetration index can be defined as the ratio of penetration depth attained per single blow, given by Eq. (4). Test results have shown that for most of these tests the penetration index was always well above 10 mm/blow mark.

This simply implies that with each and every blow, the cone could pierce through the material to depth more than 10 mm, hence, meaning that the material is weak and could easily be penetrated by the DCP. The penetration indices specifically ranged at figures greater than 10 mm/blow and less than 100 mm/blow for the majority of the tests. In addition, a trend could be established with these indices as it was with the bearing pressures. For depths of 100 mm to about 400 mm, the penetration index values are high. From depths of 500 mm and above, these values dropped sharply moving towards the 10 mm/blow mark. The drop in these was attributed to partial increase of in situ material's consolidation.

$$\text{Penetration Index} = (\text{Depth of Penetration})/(\#\,\text{of Blows}) \tag{4}$$

Averages were determined for each test data, as indicated in Table 3, through summing up the penetration indices figures and dividing by their respective numbers. The averages range between 24 and 30, with two "anomalous" figures of 58 and 16 for test points T2 and T3, respectively. Figure 1 is a sample of the graphical expressions of penetration indices taken from test point T3. The figure represents a plot of penetration index, in mm/blow, versus depth of penetration in mm.

4.3 Tailings Bearing Capacities

In situ CBR values were determined prior to extrapolation of material's bearing capacities. The PI values accompanied in situ CBR values which were acquired via calculations made on each blow taken with the DCP for each test. The formula used on this scenario is indicated by Eq. (5). These CBR values implicate the material's strength or stiffness under the prevailing in situ conditions.

$$\mathrm{CBR} = (10)^{((2.465-(1.12\times \mathrm{Log(PI)}))}. \tag{5}$$

The calculated in situ CBRs for all tests range from 02 and 05 to 14 and 22. One "outlier" value of 227 existed for test point T5, although all the other test results revealed maximum CBR at lower values. A general trend was also observed with the CBR data, with shallow depth values showing lower ratios and a considerable increase of CBR ratios with depth. Table 4 shows minimum, maximum, as well as average CBR values for each test. CBR averages for all test points, except for test points T2 and T5, were between values of 08 and 12. T2 was at 03 and T5 at 41. Nevertheless, majority of the tests proved that the material exhibited very low CBR values.

In every test, each CBR value for every blow was used to extrapolate the material's bearing capacities, using Eq. (6). These bearing capacities essentially indicate the material's capability to withstand exerted pressure and amount of shear forces.

$$\text{Bearing Capacity} = (17.58 \times (\mathrm{CBR}))^{64}. \tag{6}$$

On further analysis of the results from these tailings material, it was also noted that the material exhibited bearing capacities, hence bearing pressures, ranging in

Table 3 Summarized penetration indices results

Test no	Minimum PI (mm/blow)	Maximum PI (mm/blow)	Averages
T1	20	60	25
T2	40	95	58
T3	15	85	25
T4	10	90	24
T5	1.5	95	16
T6	20	40	30
T7	20	65	29

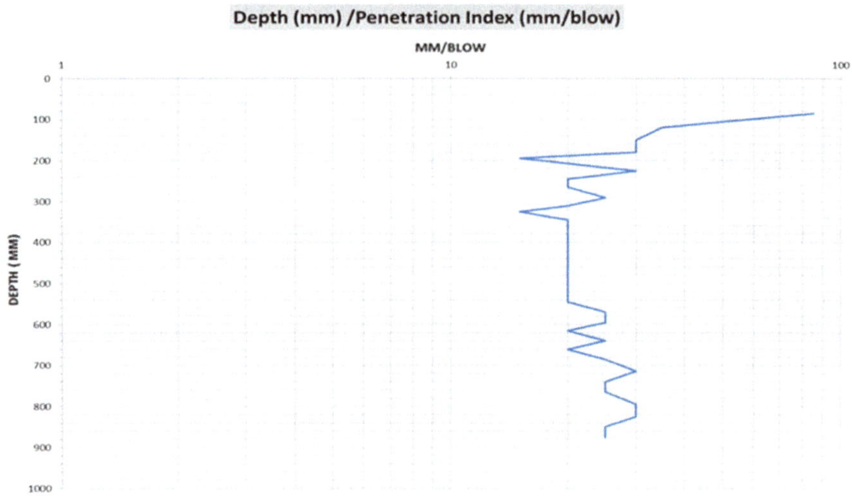

Fig. 1 Graphical expression of penetration index at a test point T3

Table 4 Summarized in situ CBR results

Test no	Minimum CBR	Maximum CBR	Averages
T1	03	10	08
T2	02	5	03
T3	02	14	09
T4	02	22	12
T5	02	227	41
T6	05	10	07
T7	03	10	07

between 20 and 100 kPa, with one test (T5) in particular showing bearing capacity as high as 500 kPa. A general trend that could also be deciphered was that the material's bearing pressure increased with increase in depth. For majority of the tests, the material showed that at depths of about 80–100 mm the bearing pressures are normally between 20 and 40 kPa. A gradual decrease in depth of testing sees the pressures take a steady rise in values. At midway through, i.e. at depths of about 500 mm, the bearing pressures are seen rising to up to 100 kPa. It is after this that bearing pressures are maintained for up to depths of nearly 800 mm when maximum reach occurs, at approximately 1 000 mm in depth.

One sample for graphical expression bearing capacity of a test T3 is shown in Fig. 2, indicating a plot of depth versus bearing capacity of material as previously discussed above. Table 5 presents summarized bearing capacity results, whereby a similar routine is also seen in the case of average values of bearing capacities, in that, these figures range from 59 to 82 kPa, bar "outliers" in T2 and T5, respectively. These two tests are an exception because T2 has an extreme low figure at 37 kPa and T5 has an extreme high value of 165 kPa in terms of average.

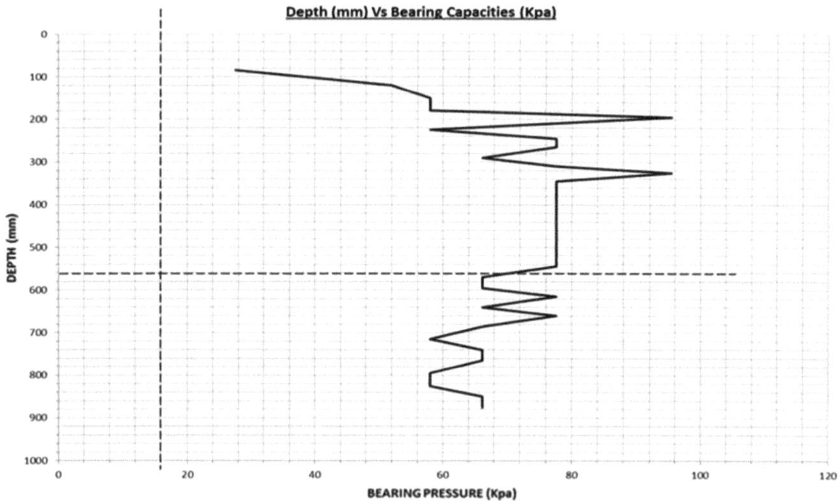

Fig. 2 Graphical expression of bearing capacity at a test point T3

Table 5 Summarized bearing capacity results

Test no	Minimum bearing capacity (kPa)	Maximum bearing capacity (kPa)	Averages
T1	35	80	68
T2	25	47	37
T3	28	96	70
T4	24	128	82
T5	20	620	165
T6	46	78	59
T7	35	80	62

4.4 Sieve Analysis Results

These results were obtained after the carrying out sieve analysis. The data shown in Table 6 indicate the per cent passing each sieve. The values clearly indicate that on the top sieve (2 mm) almost all material passed, whilst for sieves in between there was a progressive decrease of percentage of material passing the particular sieve. Sample 01 and 03 were taken along the tailings dam wall and thus exhibit a coarser fraction of the representative sample.

On the other hand, sample 08 was taken further inwards the tailings dam and indicated that material was a finer representative sample as compared to the other samples. Therefore, the amount of material passing sieve 0.075 for samples 01 and 03 is lower than that of sample 08. This is attributed to the phenomenon of

Table 6 Percent passing each sieve for all tests

Sieve size (mm)	Percentage passing (%)		
	S01	S03	S08
2	100	99.28	100
1	96.40	88.80	98.52
0.5	71.74	53.02	96.57
0.25	32.70	20.85	70.28
0.125	9.91	7.44	29.66
0.075	3.31	3.81	11.60
0.063	1.76	2.73	8.40

sedimentation, which allows for coarser material to be deposited first upon disposal of the tailings and the finer material to be deposited later further inside the tailings dam. Nonetheless, a trend is noticeable in that less fraction of material passed last two bottom sieves.

This data were then used to plot the particle-size distribution curves as shown in Fig. 3. These curves are essentially plots of sieve diameters versus percentage passing. This curve shows that the material is poorly graded, with the vital gravel-sized particles missing from the material. Much of the material's particles can be classified as clay, silt and the remaining portion makes up the sand-sized particles. Such is also evident from the data that is represented by Table 7, which shows that a bigger portion of the material is retained by sieves between the 0.25 and 0.075 mm diameter range.

The gradation test curves on Fig. 3 further validates that the tailings particles distribution covers parts of clay, silt and sand-size particles. Overall, the material

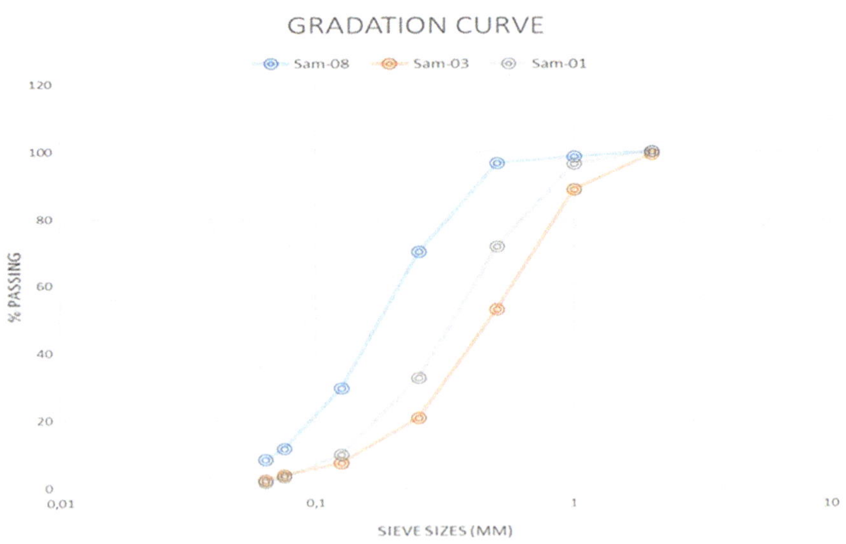

Fig. 3 Combined particle-size distribution curves

Table 7 Sieve analysis data on percent passing each sieve

Sieve size (mm)	Percentage of material retained (g)		
	Test 1	Test 2	Test 3
	S01	S03	S08
2	0	0.72	0
1	3.86	10.48	1.48
0.5	24.40	35.78	1.95
0.25	39.04	32.17	26.29
0.125	22.79	13.41	40.62
0.075	6.60	3.63	18.06
0.063	1.55	1.09	3.20
Pan	1.76	2.72	8.40
Total retained weight	576.34	644.97	620.85
Original weight	585.60	650.00	627.35
Weight loss	9.26	5.03	6.50

consists of particle sizes at a range of 0.07–1.2 mm particle diameter range, although with much of the material falling into the 0.15 and 0.8 mm zone.

Essentially, the gradation test results show that the material consists, on average, 6% fines and the remainder being sand-sized particles, thus rendering it poorly graded. The abundance of uniform sand-sized particles impacts negatively on the mechanical behaviour, and therefore reuse opportunities for engineering purpose of the material, taking into consideration issues of bearing capacity, shear strength, void ratio and so forth.

Effective size (D_{10}) for respective curves was extracted. Effective particle sizes are those particle diameters corresponding to 10% finer. D_{30} and D_{60} which are particle diameters corresponding to 30 and 60% finer, respectively, were also extracted from the curves. The resultant data were used to compute the two other crucial geotechnical parameters of material, the coefficient of uniformity (CU) and coefficient of curvature (CC), with the use of Eqs. (2) and (3), respectively. The summary of findings is presented in Table 8.

The data presented in Table 9 about per cent passing each sieve were used to give the material a group as per the AASHTO classification system. The summary of these findings is presented in Table 9 which shows that the material, on average, has between 3 and 12% fine-grained particles and roughly 90% sand-sized particles.

Table 8 Summarized interpretation of gradation results

Sample no	Effective size	Coefficient of uniformity (C_U)	Coefficient of curvature (C_C)
S01	0.13	3	1.2
S03	0.15	3	1.1
S08	0.07	3	1
Average	0.12	3	1.1

Table 9 Group classification of material using AASHTO system

Test number	Sample number	Gravel (%)	Sand (%)	Silt/ clay (%)	AASHTO class
T1	S08	0	88.4	11.6	A-3
T2	S03	0	96.19	3.81	A-3
T3	S01	0	96.69	3.31	A-3
Average		0	93.75	6.25	A-3

5 Conclusions

High PI values mean that tailings material's penetration resistance, strength and stiffness are low. The cone of the DCP could pierce through the material with ease. Therefore, it can be deduced that the material is of poor mechanical strength in this respect. The low CBR and bearing capacity values attained by the material are evidence that the material's bearing strength is questionable. This makes its reuse probably as subgrade, foundation fill and other engineering works impracticable. With coefficient of uniformity less than 3, it is concluded that the material is poorly graded. All materials with CU that is less than 4 are considered poorly graded in geotechnical terms. Therefore, crucial gravel-sized particles are missing from the material, plus almost uniform particle constituency are factors which condemns material to poor mechanical behaviour. The material is classified under A-3 group as per AASHTO classification system. Materials in this group are regarded as ones with good mechanical properties. This implies that even though the material is of poor strength; however, it still processes some favourable technical characteristics chiefly due to its low clay/silt size particle constituent. Overall, these tailings were classified as weak material according to the various mechanical properties and as such can be characterized as one with poor mechanical strength. Hence, the material's strength capabilities are doubtful and can therefore not be relied on to withstand great pressures/loads.

Acknowledgements The authors would like to acknowledge the support provided by the University of Venda, department of Mining and Environmental Geology, as well as the sponsorship awarded to the first author by National research Fund of South Africa.

References

Ahmari, S.: Recycling and reuse of wastes as construction material through geopolymerization. A Dissertation Submitted to the Faculty of Civil Engineering and Engineering Mechanics. University of Arizona, pp. 20–35 (2012)

Das, B.M.: Advanced Soil Mechanics, 2nd edn, p. 457. Taylor & Francis Ltd., London (1997)

Das, B.M.: Fundamentals of Geotechnical Engineering, 4th edn. B.H.C. Sutton. Solving Problems in: Soil Mechanics 2nd edn., pp. 753 (2006)

Fang, H., Daniels, J.L.: Introductory Geotechnical Engineering: An Environmental Perspective. Tailor and Francis, London, New York (2006)

Holtz, R.D., Kovacs, W.D.: An Introduction to Geotechnical Engineering, pp. 55–60. Prentice Hall, Inc., New Jersey. ISBN 0-13-484394-0 (1981)

Murthy, V.N.S.: Geotechnical Engineering: Principles and Practices of Soil Mechanics and Foundation Engineering. Civil and Environmental Engineering. CRC Press, New Delhi. ISBN 10:0824708733, pp. 70–80 (2002)

USEPA.: Technical Report: Design and Evaluation of Tailings Dams. U.S. Environmental Protection Agency, Washington, DC. EPA530-R-94-038. NTISPB94-201855, pp. 3–4 (1994)

Zein, A.K.M.: Use and applications of the static cone penetration test (cpt) method for the characterisation and prediction of local soils behaviour; the building and road research institute (BRRI) experience. Building and Road Research Institute (BRRI), University of Khartoum, pp. 8–9 (2007)

Part V
Water and Effluents: Treatment and Management

Assessment of Groundwater Quality: Case Study of Tshivhasa, Limpopo Province, South Africa

F. Sengani and T. Zvarivadza

1 Introduction

The investigation was conducted in the following villages: Tshikombani, Milaboni, Tshirenzheni, Mudunungu, and Dopeni, which form a part of Tshivhase area within the Thulamela Local Municipality of Vhembe District in the Limpopo Province of South Africa. Tshivhase is located in the western part of Thohoyandou Central Business District (CBD). The required water of this area is provided from a several number of wells near the villages and also from dams around the area (see Fig. 1). According to Loxton et al. (1969), the Tshivhase areas are underlain by Precambrian basalts of the Sibasa Formation of the Soutpansberg Group to the north and leucocratic biotite gneiss, leucocratic granite and pegmatite, gray biotite gneiss and migmatite of the Sand River Gneiss of the Central Zone of the Limpopo Belt to the south.

1.1 Previous Studies on Groundwater Quality

Groundwater is the water stored underground in rock fractures and pores that lay beneath the surface of the earth (AWA 2007). Freeze and Cherry (1979) outlined that groundwater is considered to be the subsurface water that occurs beneath the water tables in soils and geologic formations that are fully saturated. The principal source of groundwater is meteoric water (precipitation from rain, snow, and hail),

F. Sengani (✉)
Department of Mining and Environmental Geology, School of Environmental Sciences, University of Venda, Thohoyandou, South Africa
e-mail: fhatugeorge@gmail.com

F. Sengani · T. Zvarivadza
School of Mining Engineering, The University of the Wiwatersrand, Johannesburg, South Africa

E. Widzyk-Capehart et al. (eds.), *Proceedings of the 18th Symposium on Environmental Issues and Waste Management in Energy and Mineral Production*,
https://doi.org/10.1007/978-3-319-99903-6_18

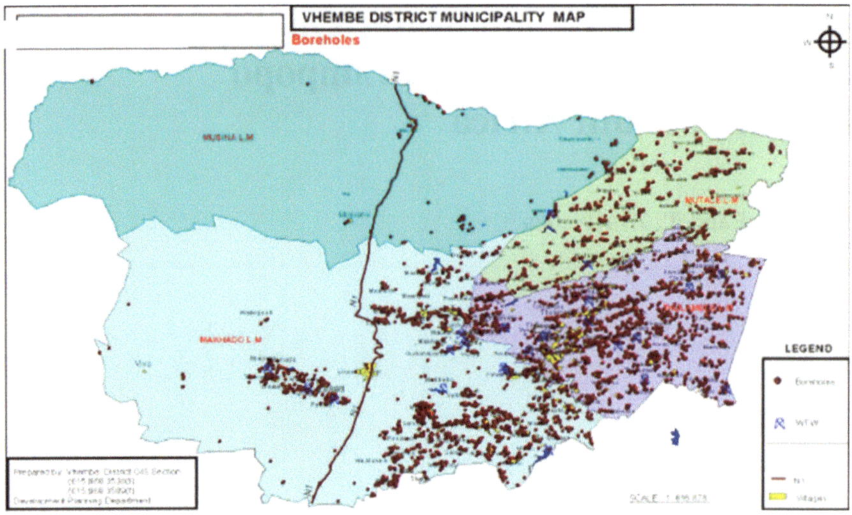

Fig. 1 Vhembe district municipality map (Vhembe municipality, 2013)

juvenile water, and connate water (Gleick 1993). According to Chapman (1996), groundwater is the most important component of the hydrological cycle, an important source of potable water in Africa, and constitutes about two thirds of the freshwater resources of the world.

Sirila et al. (2010) pointed out that water aquifers are large in extent (1–10 km) and have variations in physical and chemical properties at small scale (1–100 m). This can cause a challenge in expecting transport from potential leakage source to the receptor. Transport of contaminants in soil is a major problem for different flow scales, from the fractured rocks to large underground aquifers (Hammon et al. 2011; Sanden and Bergstrom 1986; Vladimir 2003). Deep consolidated formations are characterized by slow groundwater movement, long residence time, ample opportunity for dissolution of minerals and therefore often poor natural water quality. These formations are confines under thick sequence of low permeability clays and are less vulnerable to anthropogenic influences (Chapman 1996; Chapman and Kimstach 1996).

The quality of drinking water is typically determined by monitoring microbial presence especially fecal coliform bacteria and physicochemical properties (Gray et al. 1994; US Environmental Protection Agency 1990). Fecal population assessment in water can be achieved by determining the number of fecal coliform that are present in the water sample.

Based on DWAF, (DWAF 1996b; Kempster et al. 1997; WHO 2004), the water monitoring parameters are used to determine if the water is suitable for human consumption, recreation and other purposes such as agricultural use. Human activities such as agricultural activities, including fertilizing the soil and septic systems, are the potential sources of groundwater pollution (Cassidy et al. 2001; Freeze and Cherry 1979; Gleick 1993; Kirsh 2008; Lyle and Raymond 1998; Todd and Mays 2005; Smithson and Acworth 2005; Sverjensky 1986). According to

McDonald and Kay, (1986), the analysis of physical parameters of groundwater such as turbidity, temperature, total dissolved solids, odor, and taste is the most important physical properties of groundwater in relation to its quality.

2 Methodology

The groundwater quality properties such as pH, electrical conductivity (EC), turbidity, temperature, TDS, and salinity were measured in the field. The pH was measured using the pH meter, and turbidity was measured using an electronic turbidity meter, while the electrical conductivity and temperature of the groundwater were determined using conductivity meter.

2.1 pH

Water samples of 60 in number, which were collected in different boreholes and also in different time zones, were used. The pH of the water samples was measured by a digital pH meter. The measurements of the pH of the water were carried out in the field and also in the laboratory. In the field, the pH meter electrodes were rinsed properly with distilled water after calibration. Then, the glass electrodes were dipped into the beaker containing the water sample and it was stirred until the reading was stabilized at a certain point. In the laboratory, the samples (water samples) were also measured, all the samples were measured in triplicate, and the mean value was considered in this investigation).

2.2 EC

The measurements of the EC of the water sample were carried out in the field and also in the laboratory. In the field, the conductivity meter electrode was rinsed properly with distilled water after calibration. In the laboratory, the samples (water samples) were measured in triplicate and the mean value was analyzed.

2.3 Turbidity

A clean dry beaker was rinsed three times with the water sample as to determine the turbidity of the water under study. The beaker was filled with the water sample under study and then covered with a light shield cap. The outer surface of the beaker was dried by wiping it with a clean tissue paper. It was then pushed firmly into the optical well and then the lid was subsequently closed. The NTU values were measured by pressing and releasing the arrow, and the value was recorded after the display had stopped flashing the turbidity meter.

2.4 *TDS and Salinity*

The TDS and salinity of the water sample were measured using a multiprobe that is used to measure TDS and salinity (ExStik/Extech instrument). The measurements of the TDS and salinity of the water were carried in the field and also in the laboratory. In the field, the TDS and salinity ExStik was rinsed properly with distilled water after calibration. In the laboratory, the samples were also measured in triplicate and the mean value was considered.

2.5 *Geostatistical Analysis*

All spatial data analyses were done using the ArcMap program within the ArcGIS 10.2 software package. The spatial interpolation of water quality parameters employed in this study is the inverse distance weighting (IDW) embedded in the Spatial Analyst extension. The IDW is a default method used to generate a surface when attribute values are available at sample locations. IDW method is intended to modify the weights with information derived from the empirical data.

3 Data Collection

Samples were collected once a month over a period of four months (that is from October 2015 to January 2016). Field measurements of pH, EC, TDS, and turbidity were done using standard methods of analysis as prescribed by South African National Standards (SANS 241, 2006). A GPS was used to locate the position of the wells in the study area.

4 Results of the Physical Parameters of the Groundwater in the Study Area

4.1 *pH*

According to DWAF (1996b, 1998), pH is a measure of the hydrogen ion H^+ within water. The pH of the water affects the solubility and availability of nutrients, and how they can be utilized by aquatic organisms. Chapman and Kimstach (1996) pointed out that the acidity of groundwater is due to the presence of organic acids in the soil as well as those of atmospheric origin infiltrated to the water. The pH status of water is one of the most common parameters that needs to be identified in order to understand the water quality of certain area; however according to DWAF standards of 1998, a minimum pH value of 6.5 to a maximum value of 8.5 is acceptable. Any values less or more than the recommended standard is considered to be dangerous and toxic for domestic use.

The results from the study areas show that the pH value of 6.61 was the lowest value that was measured among all the boreholes while a pH value of 8.26 was the highest of them all. The results of the pH test are clearly illustrated by Fig. 2, wherein red solid line indicates both maximum and minimum pH values recommended according to DWAF standards. Looking at pH test results of all the wells, it was concluded that groundwater within these areas consist of pH values are within the recommended standard, however, is not enough to reach a conclusive statement on the quality of water since there are still more parameters need to be taken into consideration.

During the four months of the investigation, spatial variation of the pH was determined using ArcGIS 10.2 software. However, the spatial variation results were similar to the results obtained from physical parameters (see Fig. 3). Figure 3 illustrates the results collected from the spatial variation of the pH within the seven wells in Tshivhase area. It was found that the pH values were ranging from minimum of 7.13 to the maximum of 8.26.

	Oct	Nov	Dec	Jan
HHB 1	7,34	7,05	7,23	7,02
HHB 2	7,55	7,04	7,15	6,95
HHB 3	7,13	6,88	6,97	7,08
HHB 4	7,59	7,82	7,84	7,9
HHB 5	8,26	6,61	6,67	6,97
HHB 6	8	7,03	7,26	6,74
HHB 7	7,55	7,12	6,84	7,05
Minimum	6,5	6,5	6,5	6,5
Maximum	8,5	8,5	8,5	8,5

Fig. 2 pH values for the wells in Tshivhase areas

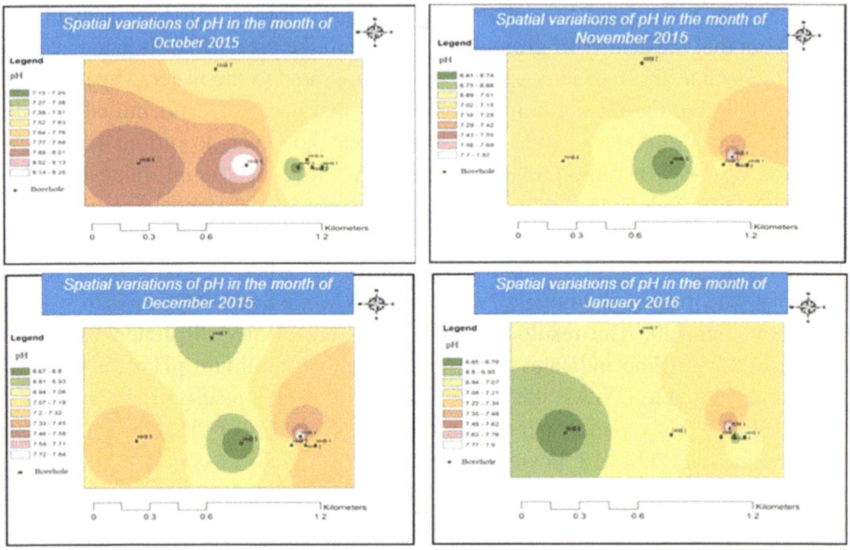

Fig. 3 Spatial variations for pH of the study areas

4.2 *Electrical Conductivity*

Electrical conductivity (EC) is a measure of the ability of water to conduct an electrical current measured in μS/cm, and this ability is a result of the presence of certain ions in water such as carbonate, bicarbonate, chloride, sulfate, nitrate, sodium, potassium, calcium, and magnesium, all of which carry an electrical charge (DWAF 1996a). If the electrical conductivity of water is high, it indicates a high concentration of ions which then determines if such water is drinkable or not (Bruvold and Ongerth 1969). The electrical conductivity of water depends on the water temperature: the higher the temperature, the higher the electrical conductivity would be (Cassidy et al. 2001).

The electric conductivity of the groundwater of the study area was also analyzed, and the results are illustrated in Fig. 4; the EC values were found to range from 945 to 2350 μS/cm. The EC values of the study area were found to be higher as compared to recommended standard of <700 μs/cm by DWAF (1998). Figure 4 shows that the EC values were above the safety level throughout the four months (from October 2015 to January 2016). The electrical conductivity of water is used as an indicator of how salt free or ion free the water sample is, the purer the water the lower the electrical conductivity and the higher the resistivity. The spatial variation of the EC of the all the boreholes was also assessed. The EC of different boreholes was found to be higher in different months as compared to the recommended standard of EC level by DWAF (see Fig. 5).

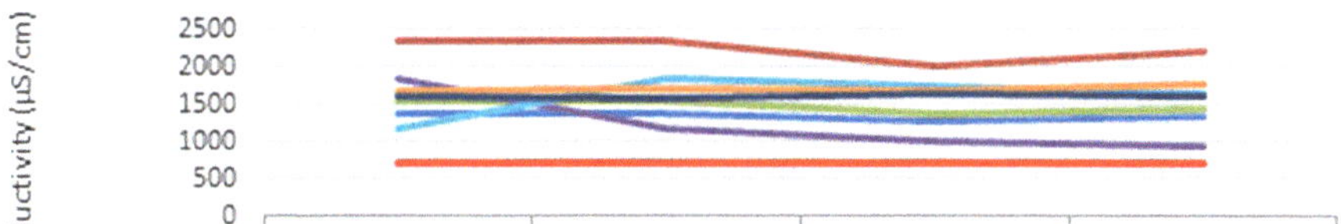

	Oct	Nov	Dec	Jan
HHB 1	1385	1387	1281	1336
HHB 2	2350	2340	2002	2210
HHB 3	1554	1535	1379	1426
HHB 4	1830	1158	1014	947
HHB 5	1191	1848	1745	1631
HHB 6	1662	1723	1673	1765
HHB7	1619	1584	1638	1604
Maximum	700	700	700	700

Time (Month)

Fig. 4 EC values for the boreholes in Tshivhase area

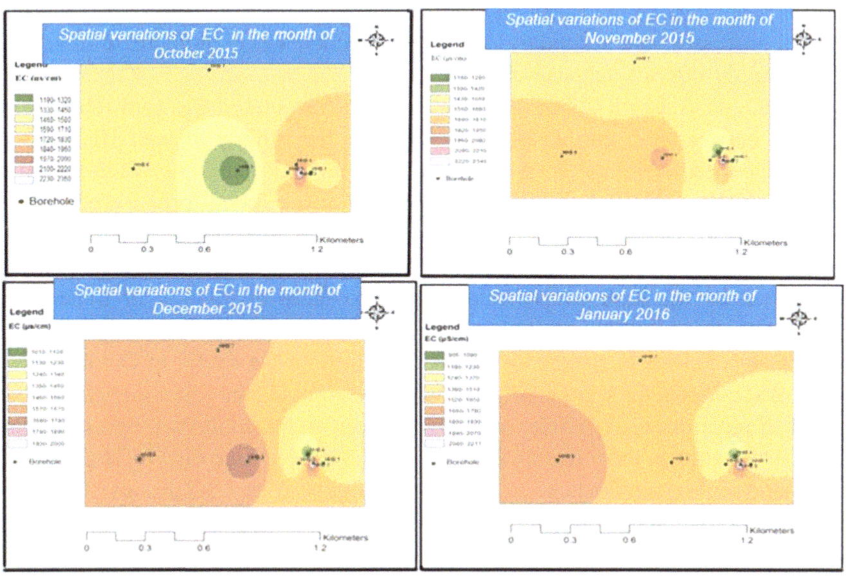

Fig. 5 Spatial variation of EC of the study areas

4.3 *Salinity*

Salinity of water samples is related to TDS concentration or electrical conductivity (Chapman 1996). The occurrence of salinity in water and soil can be divided in two groups: primary salinity, where an increase in salinity has occurred solely through

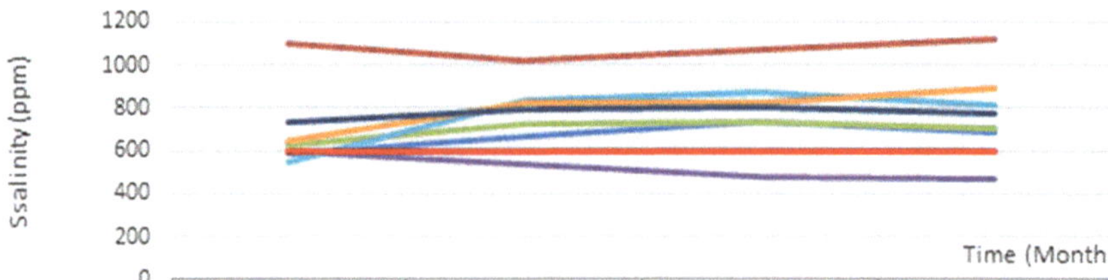

	Oct	Nov	Dec	Jan
HHB 1	588	664	732	680
HHB 2	1100	1020	1070	1120
HHB 3	628	726	732	701
HHB 4	597	537	481	465
HHB 5	547	829	869	815
HHB 6	646	824	821	891
HHB 7	731	793	803	774
Maximum	600	600	600	600

Fig. 6 Salinity values for the borehole in Tshivhase areas

natural processes, and secondary or induced salinity, where an increase in salinity has occurred due to land use changes made by human activities (Omami 2005; Smithson and Acworth 2005). Salinity values of the study area were found to be ranging between 465 and 1120 ppm. Although the salinity of well HHB 4 was within the safety drinking level which is less than 600 ppm, other wells (HHB 1, HHB 2, HHB 3, HHB 5, HHB 6, and HHB 7) were found to be exceeding the maximum safe value according to DWAF guideline (see Fig. 6). DWAF (1998), pointed out that the fluctuation in salinity value of water samples could be possibly influenced by the high temperature leading to more dissolution of minerals in groundwater.

Figure 7 shows the spatial variations of salinity in the study. In October 2015, most of the boreholes were associated with low to moderate salinity level as compared to recommended salinity values; however as the analysis progressed from October to January, there was an increase in salinity level (see Fig. 7). High salinity in water can also be influenced by high concentration of ions and dissolved minerals in the water. Salinity can occur naturally when drainage is poor or in areas with low rainfall and high evaporation, and in this case, the study area receives low rainfall.

4.4 Total Dissolved Solids (TDS)

Total dissolved solids (TDS) is defined as the concentration of all dissolved minerals in the water (Agbaire and Oyibo 2009). TDS in groundwater is due to minerals from acids produced as byproducts of the degradation process; hence, TDS is a geochemical parameter that closely links the bulk conductivity to microbial degradation

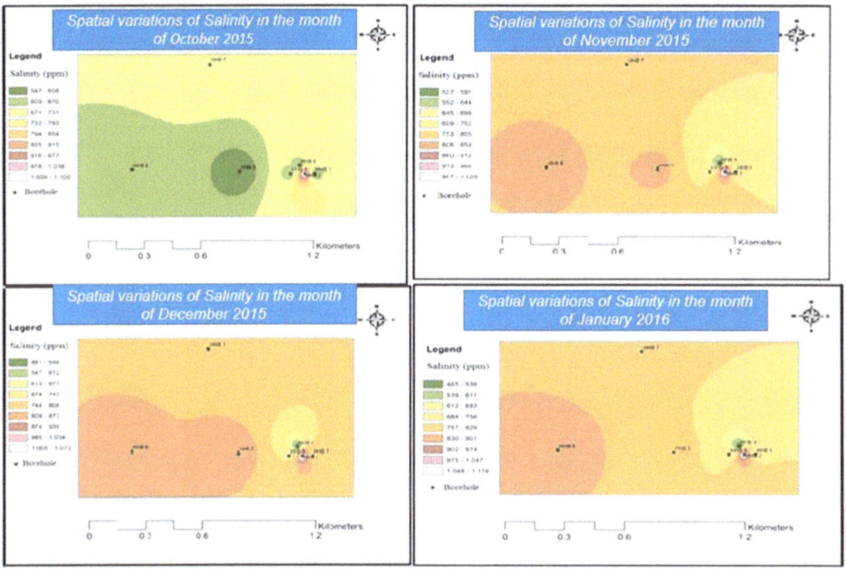

Fig. 7 Spatial variations of salinity of the study area

of hydrocarbon (Atekwanna et al. 2004; AWA 2007). TDS can be measured by automated meters or the gravimetrical method. The TDS is directly proportional to the EC of water. Since EC is much easier to measure than TDS, it is routinely used as an estimate of the TDS concentration (AWA 2007; DWAF 1996b, 1998).

Figure 8 indicates TDS of the samples for four months from October 2015 to January 2016, and the TDS values were found to be ranging from 701 to 1790 mg/l. According to DWAF standards of 1998, the maximum recommended value for TDS is 1000 mg/l. Looking at Fig. 8, it is clearly shown that most of the boreholes were found to have high TDS values as compared to DWAF standards, although the TDS at HHB1 was found to comply but in average the boreholes within the study area consisted of high TDS (see Fig. 8—left-hand side).

4.5 *Turbidity*

Turbidity is the measure of clarity in water and it is measured in Nephelometric Turbidity Units (NTU) (AWA 2007; Bruvold and Ongerth 1969; DWAF 1996b; 1998). Turbidity, which can make water appear cloudy or muddy, is caused by the presence of suspended and dissolved matter, such as clay, silt, finely divided organic matter, plankton, and other microscopic organisms, organic acids, and dyes (ASTM International 2003). The color of water, whether resulting from dissolved compounds or suspended particles, can affect turbidity measurements.

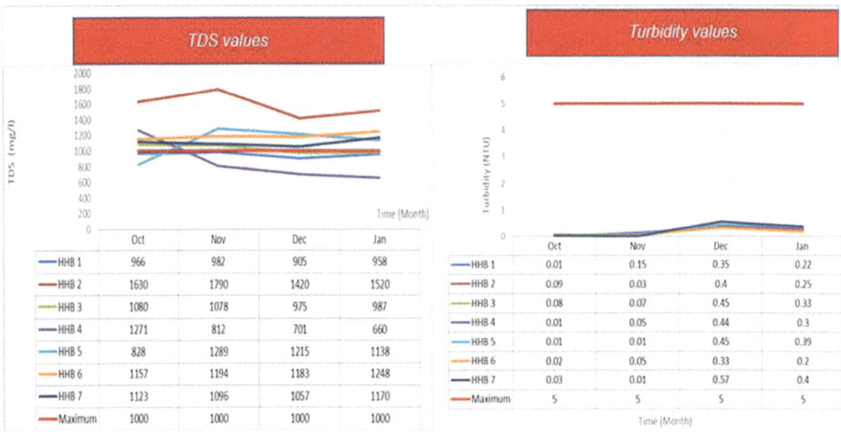

Fig. 8 TDS and turbidity values for the borehole in Tshivhase areas

Figure 8 (right-hand side) shows the turbidity values found for the boreholes in Tshivhasa areas. The turbidity values were found to range from 0.01 to 0.57 NTU (see Fig. 8—right-hand side). However, the turbidity of the study was found to be acceptable since it was within the recommended turbidity by DWAF, which ranged from 0 to 5 NTU Boreholes HHB 1-HHB7 in October 2015 were found to contain turbidity values that range from 0.01 to 0.09 NTU, in November 2015 boreholes were found to have a turbidity values that ranged from 0.01 to 0.15 NTU, in December 2015 the turbidity of the investigated boreholes were found to range from 0.33 to 0.57 NTU, and lastly, in January 2016 the turbidity values of the investigated boreholes were found to range from 0.2 to 0.4 NTU (see Fig. 8—right-hand side).

Based on the results found, it was clearly shown that the turbidity of the study area in December 2015 was relatively higher as compared to other months due to increase in precipitation. Although it was higher, it was considered to be less as per the recommended values by DWAF. However, these values clearly tell us that the groundwater within the study area was containing less microbiological contamination.

5 Discussion

After a detailed analysis of the water quality parameters (pH, EC, TDS, turbidity, and salinity) of groundwater in the Tshivhasa area, the pH values for the boreholes monitored from October 2015 to January 2016 were found to range from the minimum 6.61 to maximum of 8.26.

The turbidity of all boreholes was found to range from the minimum of 0.01 NTU to the maximum of 0.57 NTU. The turbidity values of all the boreholes were found to be within the recommended standards, which is between 1 and 5 NTU. According to WHO (WHO 2004), water quality standards, water (surface/ground) with turbidity of less than 1 NTU, are recommended for preparation of food and for drinking; however, turbidity only cannot be used to conclude on the quality of water. For drinking purpose or for food preparation, it is advisable to boil such water before drinking as way of killing bacterial and other contaminations.

The salinities of boreholes were found to range from the minimum of 465 ppm to a maximum of 1120 ppm; however, the majority of the boreholes were found to be highly saline in accordance with the recommended standards. The EC of the wells was found to be ranging from 947 to 2350 μS/cm. The EC of all the wells was then found to be exceeding the recommended standards which are less than 700 μS/cm. Lastly, the TDS values of the wells were found to range from 701 to 1790 mg/l; however, the majority of the wells in different months were found to exceed the recommended standards by DWAF.

6 Conclusions

An assessment of groundwater quality for domestic consumption was evaluated by comparing the physical parameters and spatial variation of the groundwater within the study areas and the prescribed standards by DWAF. The pH values for all the boreholes were found to be within the recommended guideline, ranging from slightly acidic to alkaline. Turbidity of the boreholes within the study area was also found to be within the recommended standard, with values that ranged between 0.01 and 0.57 NTU. The average EC, TDS, and salinity values were found to be above the recommended guideline, which clearly tells us that the groundwater is slightly salty. It was then concluded that groundwater in the study area is not suitable for domestic use but it can be used for agricultural purpose.

References

Agbaire, P.O., Oyibo, P.I.: Seasonal variation of some physico-chemical properties of borehole water in Abraka, Nigeria. Afr. J. Pure Appl. Chem. **3**(6), 116–118 (2009)

ASTM International: D1889-00 Standard test method for turbidity of water. In: ASTM-International, Annual Book of ASTM Standards, Water and Environmental Technology, West Constrohocken, Pennsylvania, vol. 11.01, 6 p (2003)

Atekwanna, E.A., Estella, A.A., Rebecca, S.R.: The Relationship of Total Dissolved Solids Measurements to Bulk Electrical Conductivity in an Aquifer Contaminated with Hydrocarbon. University of Missouri-Rolla, USA (2004)

AWA: Water in Australia: Facts and Figures, Myths and Ideas. Australian Water Association, Sydney. (2007), http://www.awa.asn.au/AM/Template.cfm?Section=Water_in Australia

Bruvold, W.H., Ongerth, H.J.: Taste quality of mineralized water. J. Am. Water Works Assoc. **61**, 120 (1969)

Cassidy, D.P., Werkema, D.D., Sauck, W.A., Atekwana, E.A., Rossbach, S., Duris, J.: The effect of LNAPL biodegradable products on electrical conductivity measurement. J. Environ. Eng. Geophys. **6**, 47–52 (2001)

Chapman, D.: Water Quality Assessments. A Guide to Use Biota, Sediments and Water in Environmental Monitoring, 2nd edn. University Press, Cambridge (1996)

Chapman, D., Kimstach, V.: Selection of Water Quality Variables. Water Quality Assessments: A Guide to the Use of Biota, Sediments and Water in Environmental Monitoring, Chapman edition, 2nd edn, p. 595. E & FN Spon, London (1996)

DWAF: South African Water Quality Guidelines, 1st edition, vol. 1. (1996a), Domestic use

DWAF: South African Water Quality Guidelines, 2nd edition, vol. 1. (1996b), Domestic use

DWAF: Quality of Domestic Water Supplies Assessment Guide, vol. 1: WRC Report No: TT.101/98, p. 101 (1998)

Freeze, R.A., Cherry, J.A.: Groundwater. Prentice-Hall, Englewood Cliffs (1979)

Gleick, P.H.: Water in Crisis. A Guide to the World's Fresh Water Resources. Pacific Institute for Studies in Development, Environment and Security Stockholm Environment Institute (1993)

Gray, N.F.: Drinking Water Quality, Problems and Solutions. Wiley, England (1994)

Hammon, H., Ginzbury, I., Boulerh cha, M.: Two relaxation-times lattice Bottzmann schemes for solute transport in unsaturated water flow, with a focus on stability. Adv. Water Resour. **34**, 779–793 (2011)

Kempster, P.L., Van Vliet, H.R., Kuhn, A.: The need for guidelines to bridge the gap between ideal drinking water quality and that quality which is practically available and acceptable. Water SA **23**(2), 167–169 (1997)

Kirsh, R.: Groundwater Geophysics: A Tool for Hyrogeophysics. Springer Science and Business Media (Illustrated Version), 568 pages (2008)

Loxton, Hunting and Associates: Natural resources survey of Venda land. Confid. Rep. TS 32/68, for Dep. Bantu Admin, Devel. S. Afr. (1969)

Lyle, S., Raymond, J.R.: Groundwater contamination. Bulletin. Water SA **2**, 3 (1998)

McDonald, A.T., Kay, D.: Water Resources Issues and Strategies. No. 3, pp. 393–404 (1986)

Murray, R., Baker, K., Revenscroft, P., Musekiwa, C., Dennis, R.: Water Research Commission Report No. 1763/1/11, Pretoria, South Africa (2012)

Omami, E.N.: Response of Amaranth to Salinity Stress. Ph.D. thesis. University of Pretoria (2005)

Sanden, P., Bergstrom, S., Gardelin, M.: Modelling Groundwater Levels and Quality. SMHI, 601 76 Norrkoping, (1986)

Smithson, A., Acworth, R.I.: An investigation of unconsolidated sedimentary units and their role in the development of salinity in the Snake Gully Catchment, Central New South Wales. The University of New South Wales Water Research Laboratory, research report No 214 (2005)

Sirila, E.R., Maxwell, R.M., Nacarre-Sitchler, A.K., McCray, J.E.: A quantitative methodology to assess the risks to human health from carbon dioxide leakage into ground water. Adv. Water Res. (2010)

Sverjensky, D.A.: Genesis of Mississippi Valley-type lead-zinc deposits. Ann. Rev. Earth Planet. Sci. **14**, 177–199 (1986)

Todd, D.K., Mays, L.W.: Pollution of groundwater. In: Jennifer, W.V., Vergas, A., Jennifer, P., Dawn, L.S., Kulesa, T., Levy, D., Morris, R (eds.) Groundwater Hydrology. Argosy publishing, Texas (Chapter 8), 361–366 (2005)

US Environmental Protection Agency: Groundwater, Volume 1: Groundwater and Contamination, EPA/ 625/6- 90/ 0160. Center for Environmental Research Information, Cincinnati, Ohio (1990)

Vladimir N.: Water Quality, Diffuse Pollution and Watershed Management, 2nd edn. Wiley, Boston (2003)

WHO: Guidelines for Drinking Water Quality: Recommendation, vol. 1, 3rd edn. World Health Organisation, Genera (2004)

Coke-Based Carbon Adsorbent

S. Yefremova, A. Kablanbekov, K. Anarbekov, L. Bunchuk, A. Terlikbayeva and A. Zharmenov

1 Introduction

It is widely known that one of the main directions for improving the efficiency of the collecting and cleaning equipment is the use of the adsorption technologies. Among a wide variety of sorbents, carbon adsorbents are characterized as the most effective ones for the removal of a wide range of adsorbates (Yahyaa et al. 2015; Chen et al. 2011). There are various sources of raw materials for the production of activated carbons. Recently, a tendency of producing them either from plant raw materials or from plant waste has been outlined (Yahyaa et al. 2015; Chen et al. 2011; Efremova 2012; Zharmenov et al. 2015; Sharma et al. 2013). However, in practice for the production of the carbon adsorbents the waste of burning solid fuel is used as well as coal mining waste, crushed coke, and activated coke (Ahmadpour and Do 1996; Zhang et al. 2017).

A technology for the special coke production from non-caking coal was developed, and a thermal-oxidation coking plant was created in the Metallurgy Center, which is a branch of the NC CPRM RK RSE. Produced special coke from Shubarkol coal is characterized by high quality as a reducing agent for metallurgical processes. However, coke grain fraction from 5 to 25 mm is suitable to these processes, while finer particles are not used and form waste. It was reported that activated coke could be used as a promising low-cost adsorbent because of its high mechanical strength and developed system of pores (Zhang et al. 2018). Due to the fact that adsorbent quality depends on the precursor's properties, there are different approaches for coke-based and coal-based carbonaceous adsorbents production.

S. Yefremova (✉) · A. Kablanbekov · K. Anarbekov · L. Bunchuk
A. Terlikbayeva · A. Zharmenov
National Center on Complex Processing of Mineral Raw Materials
of the Republic of Kazakhstan, Almaty, Kazakhstan
e-mail: s_yefremova@cmrp.kz; secretar_rgp@mail.ru

E. Widzyk-Capehart et al. (eds.), *Proceedings of the 18th Symposium on Environmental Issues and Waste Management in Energy and Mineral Production*,
https://doi.org/10.1007/978-3-319-99903-6_19

The aim of the current work is to test special coke fines forming in the process of the Shubarkol coal coking as a raw material to produce an effective adsorbent for removing rare metals from aqueous media.

2 Materials and Methods

2.1 Characterization of Special Coke

For the present research, special coke produced in the Metallurgy Center (East Kazakhstan Region) on the thermal-oxidation coking plant of coal was used.

The special coke was characterized by the following technical characteristics: humidity—1.5–2%; ash content—6–9%; and volatile—9–12%.

Grain-size distribution of the received batch of the special coke is mainly presented by the fines up to 10 mm (54.6%), with the class 0–2 mm accounting for 14.6%, class 2–5 mm—11.9%, and class 5–10—28.1%, and 45.4% is assigned to the class more than 10 mm.

The fraction of 2–5-mm fineness as special coke fines was used in experiments.

X-ray diffraction analysis (XRDA), scanning electron microscopy (SEM), transmission electron microscopy (TEM) were performed to study phase composition and structure of coke fines.

SEM and X-ray spectral microanalysis were performed on the micro-analyzer Superprobe 733 (JEOL, Japan). The analyses of the element compound of the samples and the photographic survey in various types of radiation were performed with the use of energy-dispersive spectrometer INCA Energy (Oxford Instruments, England), installed on the electron probe micro-analyzer Superprobe 733, with the accelerating voltage of 25 kW and the survey current of 25 nA. To prevent the formation of the charge at the analyzed material, able to deflect the electron beam, the samples were preliminarily coated with the thin structureless film of gold in the facility of the ionic sputtering FINE COATE (Japan). Photographs were made in the secondary electrons.

The TEM was carried out with the transmission electron microscope JEM 100CX (Japan) at the accelerating voltage 100 kV within the range of the electron microscopic magnification 9–230 thousand times. For exposure, the samples powdered in the agate mortar were placed on the object copper mesh pre-coated with the substrate film of the amorphous carbon and fixed in the microscope object holder.

XRDA was performed on the automated diffractometer DRON-3 with CuK_{α}-radiation, β-filter. Parameters of the XRD patterns shooting: U = 35 kW; I = 20 mA; shooting $\theta - 2\theta$; detector 2 °/min. X-ray phase analysis on the semiquantitative base was performed on XRD patterns of the powder samples with the application of the method of equal weighed samples and artificial mixtures. The quantitative correlations of crystal and carbon-containing phases were defined in accordance with a published procedure (Efremova et al. 2016). The interpretation of the XRD patterns was performed using the ICDD card index data: base of the powdered diffractometric data PDF2 (Powder Diffraction File) and the XRD patterns of the clean minerals.

2.2 Carbon Adsorbents Preparation

Carbon adsorbents have been prepared in a few steps. Firstly, the separation of the special coke fines into carbon and mineral parts was carried out by the method of elutriation. For this, the special coke fines were stirred in a big amount of water in a glass tank for 30 s using a stirring rate of 200 rpm. The ratio of solid: liquid was 5 g:150 cm^3. Then, the pulp was kept at rest. Due to the fact that the particles of the mineral part are much heavier than the carbon particles, at the quiescence state of the pulp they were settling on the bottom. The carbon material was staying on the surface of the water. In order to achieve the crossover, kerosene was added to the water at the ratio of kerosene: water = 1:8. Next, obtained carbon materials were subjected to activation.

Activation of the carbon materials as well as special coke fines was carried out with live water steam in two stages. The first stage was carried out at 850 °C for 1 h. After cooling, the activated material was heated again to 850 °C and was subjected to re-activation with live water steam for 1 h, while both stages of activation water consumption were changed from ~9 to ~25 kg per 1 kg of the adsorbent.

2.3 Adsorption Studies

The adsorptive properties of tested samples were characterized by the index of adsorption capacity toward iodine and rare metals.

The iodine adsorption capacity (A_i) was determined using the conventional method of the difference in iodine content in the solution before and after its interaction with the adsorbent and was calculated from:

$$X = \frac{(V_1 - V_2) \cdot 0.0127 \cdot 100 \cdot 100\%}{10m}, \quad (1)$$

where V_1 is the volume of the sodium thiosulphate solution with concentration of exactly 0.1 mol/dm^3 (0.1 N), used up for the titration of 10 cm^3 of iodine solution in potassium iodide, cm^3; V_2 is the volume of sodium thiosulphate solution with concentration of exactly 0.1 mol/dm^3 (0.1 N), used up for the titration of 10 cm^3 of iodine solution in potassium iodide after its contact with the adsorbent, cm^3; *0.0127* is the iodine mass relevant to 1 cm^3 of sodium thiosulphate solution with concentration of exactly 0.1 mol/dm^3 (0.1 N), g; *100* is the volume of iodine solution in potassium iodide taken for clarifying with the adsorbent, cm^3; *10* is the volume of iodine solution in potassium iodide taken for titration, cm^3; and *m* is the mass of adsorbent used, g.

The adsorption of rare metals was studied using the model ammonium perrhenate solution with concentration of Re(VII) as 35.8 mg/g. Prepared solution had the pH value of ca. 6 that was measured using EV-74 ionomer (Plant instrumentation, Byelorussia). For adsorption studies, 1 g of investigated adsorbents was mixed with 50 cm^3 of the prepared solution in Erlenmeyer flasks and shaken in the rotary shaker at 150 rpm at the room temperature for 2 h. Next, the mixtures were filtered and the concentrations of rhenium (VII) ions in the filtrates were analyzed using Agilent AA240FS atomic absorption spectrometry (Agilent Technologies, USA).

3 Results and Discussion

3.1 *Structure and Morphology of the Special Coke Fines*

According to SEM results, special coke fines are formed by the particles of various sizes with different structures (Fig. 1). The main mass is represented by non-porous particles with the size of up to ~40 μm; monolithic lumps having the size of up to 150 μm with smooth and shell-like surface occur as well. There are quite coarse particles (up to 500 μm) that like charcoal are characterized by an uneven structure in different crystallographic directions. The layered structure allows referring to this type of microstructure to a lamellar one. There are some coarse–porous particles with a pore size of 50–100 μm. They have thick, dense, smooth, melted contours of the walls. There are particles with the developed pore system. The size of the rounded pores varies from less than 1 to 30 μm. Slit-shaped pores are also observed, as well as pores elongated in the oval form which are up to 20-μm diameter and up to 40 μm long.

Using the method of X-ray spectral microanalysis, the predomination of carbon (87.5%) was established in the composition of the studied special coke fines sample. The remainder is formed by oxygen and inorganic impurities.

It was determined using TEM that the main mass of the special coke fines was formed by loose particles with different densities (Fig. 2a). The pores of the size from 2 to 5 nm that are located between the chaotically placed grains having the size of 3–7 nm are common for them. The loose particles mostly have the structure which is similar to the three-dimensional-ordered carbon. The micro-diffraction pattern from such particles is represented by rings with interlayer distances set of 0.34–0.37 nm; 0.205–0.22 nm; and 0.10–0.13 nm (Fig. 2b). Rounded flattened particles (Fig. 2a) on the weak diffuse rings on the micro-diffraction pattern (Fig. 2c) can be referred to the amorphous carbon matter.

Fig. 1 SEM micrograph of the special coke fines sample

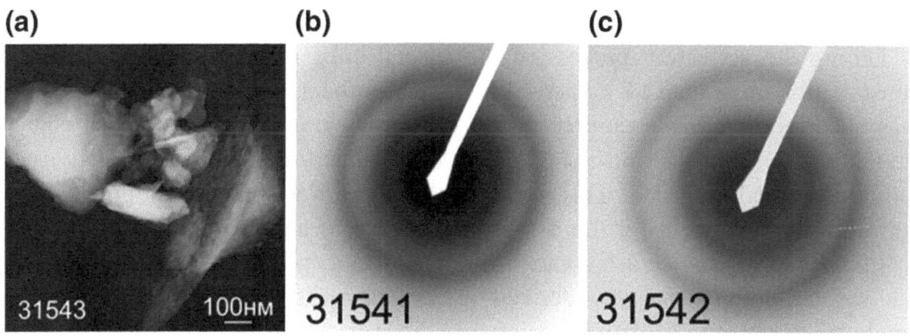

Fig. 2 Electron microphotograph and micro-diffraction patterns of the special coke fines sample

It was determined by XRDA that investigated sample of coke fines consists of carbonaceous phase, quartz, and calcite. Carbonaceous phase is formed by a graphite-like phase (G_{ph1}, 41.6%) with d_{002} = 0.3590 nm, the second graphite-like phase (G_{ph2}, 29.5%) with d_{002} = 0.3350 nm (that well conforms with the TEM data), a polynaphthenic phase (N_{ph}, 18.5%) with $d \sim 0.47$ nm, and a hydrocarbon oxygen-containing phase (H_{ph}, 10.4%) with $d \sim 0.8$ nm.

3.2 The Adsorption Activity of the Special Coke Fines and Prepared Coke-Based Carbon Adsorbents

Due to the prevailing of the non-porous particles, the particles having a graphite-like structure, as well as the particles with big pores, the studied special coke fines sample is characterized by the low index of adsorption activity on iodine (1.9%). At the same time, the pore system formed during the special coke production process is the promising factor of the micro-porosity development by the method of steam-gas activation with live water steam as an activating agent. As a result of the interaction of the special coke fines with the superheated water steam, the thermal-oxidative activation of the carbon surface takes place with emission of the carbonic oxides, hydrogen, and methane on the main and side reactions (2–5), which allows increasing and loosening of the surface and, correspondingly, improving of the sorption properties:

$$C + H_2O \rightarrow CO + H_2 - Q, \tag{2}$$

$$CO + H_2O \rightarrow CO_2 + H_2 + Q, \tag{3}$$

$$C + CO_2 \rightarrow 2CO - Q, \tag{4}$$

$$C + 2H_2 \rightarrow CH_4 - Q, \tag{5}$$

During the activation process of the special coke fines with the water steam consumption more than 14 kg per 1 kg of the adsorbent, a high degree of scorching was observed that was accompanied by the low yield (44%) and high ash content (more than 20%) of the finished product. At the same time, the sorption activity on iodine increased to 20.3%. This indicated that there was some development of the surface, but insufficient for the use of the material as an adsorbent.

In order to reduce the ash content of the produced special coke-based adsorbent, the carbon component was derived as described in 2.2. The results of the single-stage separation of the special coke fines in water and technical characteristics of the derived products are presented in Table 1, from which it ensued that the yield of the floating fraction (the carbon material) makes up 60%. The obtained carbon material (CSC-H_2O) had the index of the sorption activity on iodine at the level of 2.5%. As the research showed, almost 24% of carbon presented in the initial special coke fines sample settled together with the mineral part probably as a result of a strong bond with it, forming the heavy fraction. The yield of the heavy fraction made up 34%. Six percentage was for the loss of the fine carbon fraction with water. In order to avoid high labor contribution, the repeated separation of the heavy fraction was not carried out. However, in order to increase the completeness of separation of the special coke fines into carbon and mineral parts the hydrophobic nature of carbon was taken into account, and kerosene was added to the water during the elutriation process. This allowed increasing of the carbon

Table 1 Results of the separation of the special coke fines sample using elutriation in water and the characteristics of the obtained products

Name of the product	Yield (%)	Ash content (%)	Humidity (%)	Volatile substances (%)	A_i (%)
Floating fraction ($CSC-H_2O$)	60	0	1.6	4.9	2.5
Heavy fraction	34	30	1.3	2.2	0
Losses	6	n/a	n/a	n/a	n/a

Table 2 Results of the two-stage steam–gas activation of the special coke fines-derived carbon materials

Name of the adsorbent sample	First stage of H_2O activation			Second stage of H_2O activation				
	Water consumption (kg/1 kg of the adsorbent)	Yield (%)	A_i (%)	Water consumption (kg/1 kg of the adsorbent)	Yield, (%)	A_i (%)	Re(VII)	
							a (mg/g)	R (%)
$CSC-H_2O-1$	8.9	72.0	32.0	9.6	65.7	42.0	–	–
$CSC-H_2O-2$	17.1	69.0	22.7	18.2	57.9	47.0	1.63	94
CSC-K	10.0	66.7	29.2	24.4	47.4	50.8	1.69	96

material (CSC-K) yield up to 92%. The yield of the mineral part made up 9.5%. Also, the losses decreased by 1.5%.

After the activation of the $CSC-H_2O$ sample with live water steam, its sorption activity on iodine rised to only 32% ($CSC-H_2O-1$, Table 2). Therefore, the activated material after its cooling was subjected to the repeated heating and activation with live water steam. The A_i value grew up to 42% as a result. In another way, the step of the activated carbon material cooling was carried out under the water steam flow. This method helped to develop the sorption activity of produced adsorbent toward iodine up to 47% ($CSC-H_2O-2$ sample, Table 2). The received data indicate the effectiveness of the alternation of the heating–cooling processes influencing the formation and development of the porous structure, specific surface area, and, accordingly, the sorption capacity of the adsorbent. The special coke fines-derived carbon by the mix of kerosene and water (CSC-K) after the second stage of activation showed the maximum sorption activity on iodine as 50.8%.

The $CSC-H_2O-2$ and CSC-K samples were tested in the process of rare metals removal from the aqueous media. The results showed they were active enough in the process of rhenium removal from the water solution of ammonium perrhenate. The index of Re (VII) adsorption was 1.63 and 1.69 mg/g at the metal uptake degree of 94 and 96%, respectively, with the $CSC-H_2O-2$ and CSC-K adsorbents. At the same time, as it can be seen from Table 2, there is a trend: The higher the value of the index of the adsorption activity of the adsorbent on iodine, the higher the value of rhenium (VII) ions adsorption achieved by it.

4 Conclusions

Thus, the possibility of producing an effective carbon adsorbent based on the special coke fines is demonstrated. It is established that due to the peculiarities of the structure of the special coke fines, and in particular high ash content, prevailing of non-porous solid particles and carbon particles of a graphite-like structure, steam–gas activation of the special coke with live water steam does not give the desired result. The necessity to separate the special coke fines into carbon and mineral parts is shown. Almost, complete separation of the special coke fines with the carbon material yield up to 92% is achieved with the use of a low-cost method of elutriation in the mixture of kerosene and water at the ratio of 1:8. The necessity to perform the steam–gas activation of the special coke fines-derived carbon by live water steam in two stages is shown. The produced adsorbent is characterized by the adsorption activity on iodine of 50.8% and removal efficiency toward rhenium (VII) ions of 96%.

Acknowledgements We acknowledge the Ministry for Investments and Development of the Republic of Kazakhstan (Program No. 0066/PTF-17) for the financial support of the current work.

References

Ahmadpour, A., Do, D.D.: The preparation of active carbons from coal by chemical and physical activation. Carbon **34**(4), 471–479 (1996)

Chen, Y., Zhu, Y., Wang, Z., Li, Y., Wang, L., Gao, X., Ma, Y., Guo, Y.: Application studies of activated carbon derived from rice husks produced by chemical-thermal process. Adv. Colloid Interface Sci. **163**, 39–52 (2011)

Efremova, S.V.: Scientific and technical solutions to the problem of utilization of waste from plant- and mineral-based industries. Russ. J. Gen. Chem. **82**(5), 963–968 (2012)

Efremova, S.V., Korolev, Y.M., Sukharnikov, Y.I., Kablanbekov, A., Anarbekov, K.K.: Structural transformations of carbon materials in the processes of preparation from plant raw materials. Solid Fuel Chem. **50**(3), 152–157 (2016)

Sharma, P.K., Tripathi, C.N., Ayub, S.: Agro and horticultural wastes as low cost adsorbents for removal of heavy metals from wastewater. Int. Refereed J. Eng. Sci. **2**(8), 18–27 (2013)

Yahyaa, M.A., Al-Qodahb, Z., Zanariah Ngaha, C.W.: Agricultural bio-waste materials as potential sustainable precursors used for activated carbon production: a review. Renew. Sustain. Energy Rev. **46,** 218–235 (2015)

Zhang, C., Chen, Z., Li, J., Guo, Y., Cheng, F.: Removal of recalcitrant organic pollutants from bio-treated coking wastewater using coal-based carbonaceous materials. Desalin. Water Treat. **88**, 75–84 (2017)

Zhang, C., Li, J., Chen, Z., Cheng, F.: Factors controlling adsorption of recalcitrant organic contaminant from bio-treated coking wastewater using lignite activated coke and coal tar-derived activated carbon. J. Chem. Technol. Biotechnol. **93**, 112–120 (2018)

Zharmenov, A., Yefremova, S., Sukharnikov, Y., Kablanbekov, A., Anarbekov, K., Yesengarayev, Y.: Synthesis and determination of heavy and rare metals adsorption capacity of carbon materials from rice hulls, cellulose and lignin. In: Musingwini, C., Rupprecht, S., Genc, B., Singhal, R.K. (eds.) Proceedings of the 23rd International Symposium on Mine Planning & Equipment Selection MPES 2015, Johannesburg (2015). pp. 903–917

Part VI
Mine Ventilation

Analysis of Ventilation System and Assessment of Hazards in the Process of Progressing Liquidation of Workings in Mine 'S'

Wacław Dziurzyński, Marek Grzywacz and Jerzy Krawczyk

1 Introduction

The basic objective of the paper is to present the plan of mine workings liquidation taking into account natural hazards and to analyse the air flows during the progressing liquidation of the mine workings system. The main problems considered are the stable operation of fans installed in ventilation shafts and the fire hazard.

The planned staged liquidation of workings resulted in the necessity to regulate the ventilation system in such a way as to maintain safety of staff working at each stage of the liquidation.

To undertake the analysis, the following data are required:

- Schedule of 'S' mine workings liquidation,
- Initial structure of the considered ventilation system,
- Values of parameters characterising the air flow in the workings system,
- Current characteristics of working fans.

The data obtained via 'in situ' measurements in the mine workings system allow to use a modern computing tool, ventilation engineer's software system ***'Ventgraph'*** (Dziurzyński et al. 2013, 2015; Pritchard 2010). Computer programs of this system are used in numerous Polish mines, and the results obtained from their application are useful for the determination of ventilation system safety objectives.

The computer database held by the mine ventilation department was used in this report. The aforementioned database is the basis for variant analysis of the

W. Dziurzyński (✉) · J. Krawczyk
The Strata Mechanics Research Institute, Polish Academy of Sciences, Kraków, Poland
e-mail: dziurzyn@img-pan.krakow.pl

M. Grzywacz
Minadores, Galerias y Tuneles SL, Asturias Espana, León, Spain

E. Widzyk-Capehart et al. (eds.), *Proceedings of the 18th Symposium on Environmental Issues and Waste Management in Energy and Mineral Production*,
https://doi.org/10.1007/978-3-319-99903-6_20

progressing liquidation of workings. Because of the fire hazard, the study pays particular attention to the aerodynamic potential (Bystroń 2001) distributions for the sequence of the stages of 'S' mine ventilation system liquidation.

2 Situation of Natural Hazards

This chapter provides general information about the ventilation system and the potential hazards.

2.1 General Description of 'S' Mine Ventilation System—Situation as of December 1999

The ventilation system of 'S' mine has three intake shafts: 'Artur Główny' No. 2; 'Artur Nowy' No. 1; 'Artur' No. 4. Fresh air flows down these shafts to the mine, spreading into four sub-systems, independent of each other, and related to the following return shafts:

1. 'Wschodni',
2. 'Centralny', liquidation in December 1999,
3. 'Zofia',
4. 'Zbyszek'.

Proper division of air to individual ventilation sub-systems was made by means of regulating door in which arrangement depends on the current situation of mining operations and will change with their relocation. In workings with rope transport equipment, the doors were opened or closed mechanically, or their automatic opening or closing was ensured, without the need for the staff to enter the transport route.

Regulating doors were built only in workings being inlets to specific ventilation sub-systems, where the rules of avoiding the door location in workings with the belt conveyors and, in particular, in places where their drives were installed, were observed.

2.2 Districts Supplied with the Descending Air Flows

For safe ventilation, the bottom of the shaft supplying a mine with fresh air should be lower than the districts of exploitation. The regions below the shaft bottom are called sub-level and require special safety considerations to avoid a flow reversal during a fire. There are no sub-level districts supplied with the descending air flows in the ventilation system of 'S' mine.

2.3 Temperature Hazard

There was no need to control elevated temperatures in KWK 'S'. The highest temperature existing in 'S' mine workings so far did not exceed 23 °C.

2.4 Ventilation Connections with Neighbouring Mines

The ventilation system of 'S' mine does not have connections with neighbouring mines.

2.5 Methane Hazard

KWK 'S' was classified as a non-methane mining plant by the decision of the Regional Mining Authority in Krakow of 05/06/1970. The methane content existing in the mine corresponds to classification of the mine as a non-methane mining plant. Despite that gas recognition was carried out on a regular basis in newly opened and mined seam parts by means of an index method.

2.6 Gas and Rock Outburst Hazard

All the mined coal seams are classified as not threatened with the gas and rock outburst. Moreover, since the beginning of coal mining in the colliery there was no gas or rock outbursts.

2.7 Dust Hazard

All coal seams mined in the colliery and workings driven in them were classified as class 'A' of coal dust powder explosion hazard based on the decision Regional Mining Authority in Tychy. The coal dust powder explosion hazard was analysed at least once a year. Because of strong water flooding and a high own humidity of seams, the originating coal dust is either hardly volatile or not volatile at all.

Table 1 Spontaneous ignition group classification

No	Seam/part/district	Spontaneous ignition coefficient Sza (°C/min)	Spontaneous ignition group	Assessment of coal liability to spontaneous ignition
1	207 p 'A'	160	V	Highly liable
2	208 p 'A'	158	V	Highly liable
3	209–210 p 'A'	148	V	Highly liable
4	209–210 p 'S'	160	V	Highly liable
5	214 p 'H'	146	V	Highly liable

2.8 *Fire Hazard*

The coal existing in seams developed now, and mined is the coal of very high liability to spontaneous ignition (group V of spontaneous ignition). This assessment was made based on the spontaneous ignition coefficient Sz^a-°C/min and requirements of the Polish classification of the seam to appropriate spontaneous ignition group as shown in Table 1.

The mine had no sealed off fire zones. All the mine panels were ventilated by independent flows. Irrespective of ventilation measures aimed at the reduction of the spontaneous combustion possibility, the unnecessary workings were tightly and carefully backfilled by hydraulic filling, the opened coal parts were quickly mined, and unnecessary workings were liquidated. Voids above the roadway support and cracks in the rock mass were filled with crylamine foams and adhesives.

Fire prevention activities, aimed at preventing the spontaneous combustion in the old goaf for mining with caving, consisted of:

- Clean extraction of coal,
- Striving for obtaining as complete as possible caving with the longwalls progress,
- Immediate tight insulation of the mined deposit part,
- Fast advance of faces,
- Applying proper direction of seam mining, related to maintenance of ventilation roadways in production districts,
- Observation of ventilation rule along the coal face, resulting in substantial reduction of air flow through old goaf,
- Liquidation of longwalls within 3 months from the end of their mining.

3 Computation of Air Distribution in the Ventilation System of 'S' Mine for the Stages of Abandonment

Computation of air distribution in the ventilation systems of the 'S' mine was based on the information of the stages of liquidation (in 2000) provided by the mine. Modifications were made with the Ventgraph modules, namely the EDTXT,

EDRYS and GRAS (Dziurzyński et al. 2013, 2015, 2017). These stages refer consecutively to individual months of 2000, starting from January and ending in September. To carry out variant computations of air distribution, a database was created, describing the initial state of mine ventilation system on 1 December 1999, which was modified to create next versions of the ventilation system structure, corresponding to individual stages of the aforementioned liquidation. For those purposes, materials delivered primarily by the mine were used, including:

- Results of ventilation measurements underground and on the surface,
- Isometric diagram of mine ventilation with marked liquidation stages,
- Main fans characteristics,
- Computer database describing the ventilation system structure and parameters of ventilation nodes and branches.

Computations of air spreading were carried out in total for 14 variants related to the state of mine ventilation system for the following periods of time:

- 01/12/1999—variant 0, initial (database for the next variants),
- 01/01/2000—variant 1,
- 01/02/2000—variant 2,
- 01/03/2000—variants 3 and 3a,
- 01/04/2000—variants 4 and 4a,
- 01/05/2000—variants 5 and 5a,
- 01/06/2000—variant 6,
- 01/07/2000—variant 7,
- 01/08/2000—variant 8,
- 01/09/2000—variants 9 and 9a.

Assumptions for each next computation variant were based on the previous variant data and were supplemented with necessary changes resulting from the scope of liquidation and requirement of proper ventilation of the entire mine. In the case of four variants, computations were carried out for two sub-variants, showing alternative ventilation solutions.

For variants 3, 4 and 5 (subversions named 3a, 4a and 5a), the alternative solution was related to the lack of or to adopting intentional unsealing of 'Wschodni' return shaft cap (because of operational reasons, the implementation can be carried out in a slightly different way, e.g. by opening another inflow connected with the ventilation channel). The unsealing is aimed at reducing the airflow load of common ventilation roadway on level 256 m towards 'Zbyszek' and 'Wschodni' return shafts and at reducing the pressure of fan WPK-3,1 at 'Wschodni' shaft, thereby its capacity affecting underground workings of this shaft sub-system. Because of a high hazard of endogenous fires, existing in the 'S' mine, the reduction of fan pressure and of potential drops occurring underground the mine is a very positive preventive action, the more so that it does not result in deterioration of ventilation state of 'Wschodni' shaft sub-system workings. For the same purpose, the mine can consider also changing the angle of baffle blades of WPK-3,1 fan; however, it seems that despite such possibility existence the unsealing will be necessary anyway.

In the case of variant 9, the alternative solution was related to the last stage of liquidation, i.e. the period of ventilating relatively a small number of workings using a distant 'Wschodni' return shaft (acc. to the mine assumptions) or 'Zbyszek' return shaft (variant 9a—suggestion of report authors) situated much closer to central shafts and workings, which will be liquidated at the very end of the aforementioned stage. From the ventilation point of view, in conditions of a high hazard of endogenous fires, the liquidation of the underground part of a mine carried out from the most distant peripheral shafts (minefield boundaries) to the central shafts seems more favourable.

Assumptions of **variants 1–9** related to a general description of mine ventilation system narrowing as a result of successive disconnecting of underground workings from the system due to their liquidation or insulation, including the list of disconnected branches (workings), are presented in Table 2. Table 2 presents also changes adopted in the field of location and the state of regulating doors use (total or partial opening or closing of regulating doors and windows in the doors), related actually to the aerodynamic resistance of ventilation air splits.

Table 2 Description of basic variant assumptions

Variant number	Description
1	Disconnection from the ventilation system of a drainage road, incline II and a ventilation roadway in the 'Zofia' shaft sub-system and 'Centralny' shaft together with workings connecting it with shafts 'Artur Główny' and 'Artur Nowy' Because of difficulty in ventilating parts of pit-bottom chambers via ventilation staple pit I towards 'Zofia' shaft, the opening of regulating doors situated on the same level in the ventilation route (branch 92) towards 'Zbyszek' shaft was assumed and free spreading of air from the aforementioned chambers towards referred to return shafts. The opening of the regulating door in branch 119 was assumed for a full ventilation use of workings situated close to the 'Zofia' shaft pit-bottom on level 190 m
2	Disconnection from the ventilation system of remaining workings on levels 260 and 350 m together with inclines I and IV assigned to the 'Zofia' shaft sub-system, and also of drift Ligia level 256 m with the area of inclines III, IV, V, XXVIII and the haulage roadway assigned to the 'Wschodni' shaft sub-system. The following changes in the ventilation regulation were adopted in relation to the liquidation works in the area of longwall 604 (in the 'Wschodni' shaft sub-system): sealing dams in incline IV (branch 311), i.e. on the way of air delivery to longwall 604, installation of dams in workings below longwall 604 (branches 327J ad 342)—within the longwall area preparation to insulate, partial unsealing of regulating doors in workings below longwall 604 (branches 558 and 559)

(continued)

Table 2 (continued)

Variant number	Description
3	Liquidation of longwall 604 and of a few adjacent workings, as well as of workings ventilated from the end of eastern drift of level 350 m together with corresponding ventilation routes towards 'Wschodni' shaft. Unchanged regulation of ventilation—as in variant 2
3a	Assumptions like for variant 3 and additionally assuming partial unsealing of 'Wschodni' pit-bank (branch 41), to such extent that the depression (value of potential) in 'Wschodni' shaft at the ventilation channel would be similar to the value from variant 2 (much lower than in variant 3)
4	Liquidation of the area of longwalls 502 and 503 in the 'Zbyszek' shaft sub-system together with drifts supplying there fresh air (sections of southern and water drift on level 350 m) and removing the return air (drift to seam 207, level 350 m, southern drift—level 260 m). Unchanged regulation of ventilation, as in variant 3
4a	Assumptions as for variant 4 and additionally assuming partial unsealing of 'Wschodni' shaft pit-bank (branch 41), as in variant 3a
5	Disconnecting 'Zofia' shaft from the ventilation system (stopping the main fan) together with chambers on level 194 m and above and a single ventilation route on this level (through western drift I). Moreover, part of chambers on level 260 m was liquidated (rectifiers chamber) and on level 350 m (KMW, mine car repair chambers and electrical and mechanical workshops)
5a	Assumptions as for variant 5 and additionally assuming partial unsealing of 'Wschodni' shaft pit-bottom (branch 41)—as in variant 3a
6	Disconnecting from the ventilation system of workings in the area of longwalls 601 and 602 together with ventilation routes to 'Wschodni' shaft and its pit-bottom (level III). Because of the above operation, a single air roadway will lead to 'Wschodni' shaft (from the drift towards 'Zbyszek' shaft). In this situation to maintain stable operation of the main fan at 'Wschodni' shaft on as low as possible pressure difference and to maintain at least minimum air inflow to workings adjacent to the 'Zbyszek' shaft on level 260 m (there is a high dependence of air flows on 'Zbyszek' and 'Wschodni' shafts on this level), the following changes in the ventilation regulation were adopted: partial unsealing of 'Wschodni' shaft pit-bank (branch 41)—as in variant 3a and opening all regulating doors in the area of 'Wschodni' shaft pit-bottom, level 256 m (branches: 213, 214, 215 and 216)
7	Disconnection of 'Artur IV' return shaft from the ventilation system to start filling it. Liquidation of workings and chambers ventilated from this shaft together with the belt and technological incline. Regulation of ventilation without changes—as in variant 6
8	Liquidation of workings on level 515 m and connecting it with level 350 m together with a section of 'Artur Nowy' shaft and also sections of the eastern and water eastern drift, level 350 m and a section of 'Zbyszek' shaft between levels 350 and 260 m. Because inflows to 'Artur Nowy' shaft will have regulating doors with windows (liquidation stage preceding the shaft filling), a strong damping of the air flow in those inflows has been assumed (branches: 67, 128, 220, 221 and 512). To ensure ventilation of trunk roadways I and II pass-by on level 350 m and of adjacent workings, partial unsealing of regulating doors in branch 282 was assumed

(continued)

Table 2 (continued)

Variant number	Description
9	'Artur Nowy' shaft (to be filled) was disconnected from the ventilation system up to the highest connection with 'Artur Główny' shaft, where a regulating door with a window was built in the vicinity of shaft (branch 67), as in variant 8. Moreover, part of chambers and workings adjacent to the shaft were liquidated on levels 260 m and 350 m together with air roadways to and on level 190 m (including ventilation staple pit I). At the same time, 'Zbyszko' shaft was subject to liquidation (stopping the main fan) as well as workings towards this shaft on levels 190 and 260 m (together with the vicinity of shaft). To ensure ventilation of the remaining chambers, of trunk roadways and of other workings towards 'Wschodni' shaft, a suggestion was made to use for this objective the air roadways passing through: mine locomotives depot chamber (branch 137)—for chambers, workings—branches with nodes 507, 506, 505 and 508—for trunk roadways
9a	In this variant instead of 'Zbyszek' shaft and related workings liquidation, an alternative solution was suggested, i.e. liquidation of 'Wschodni' shaft (stopping the main fan) together with the vicinity of shaft and the ventilation route from this shaft up to the drift towards the 'Zbyszek' shaft pit-bottom Because of air delivery to 'Zbyszek' shaft by a ventilation route without air flow damping, regulating doors opening was assumed on the remaining pass-bys of the above shaft pit-bottom on level 260 m, (branches: 181, 171 and 172)

Results of air flow computations for 14 variants have been specified in the files of data sets and of computations using the Ventgraph software, while selected parameters are characterised in Figs. 1, 2 and 3 and development of potential values at selected ventilation nodes for sub-systems of shafts: 'Zofia', 'Zbyszek' and 'Wschodni' is shown in Fig. 3. Fig. 3 shows a graph of the aerodynamic potential changes for selected nodes.

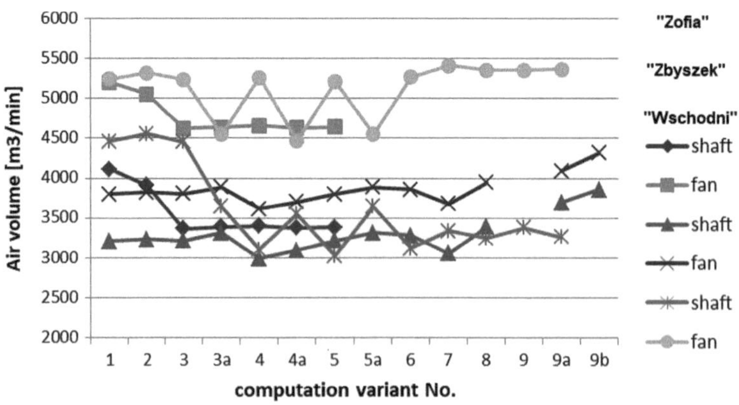

Fig. 1 Air volume in sub-systems of return shafts 'Zofia', 'Zbyszek' and 'Wschodni'

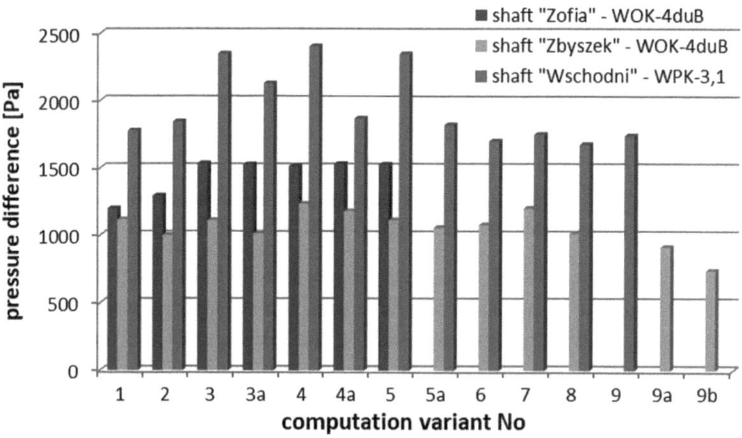

Fig. 2 Main fans pressure difference

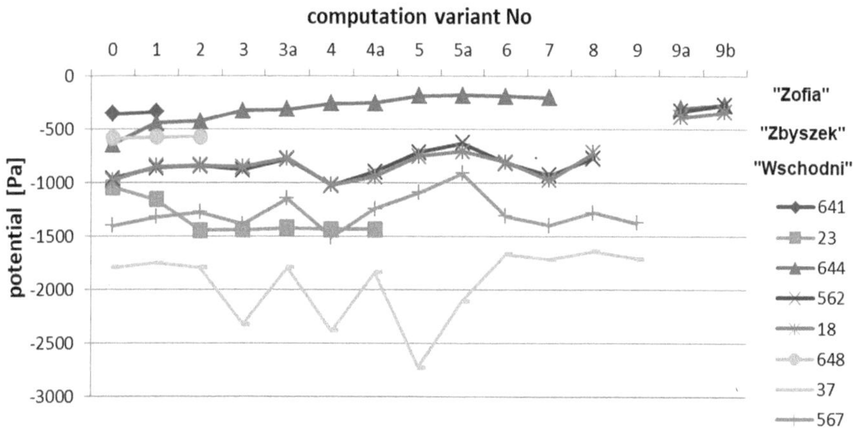

Fig. 3 Development of potential values at selected ventilation nodes for sub-systems of shafts: 'Zofia', 'Zbyszek' and 'Wschodni'

4 Recommendations

During the implementation of individual stages of mine underground liquidation, in the field of ventilation changes it might be necessary to verify certain assumptions given in the description of calculated variants, taking into account the need to ensure stable operation of main fans and proper conditions of mine workings ventilation.

Among other things, it is necessary to consider a possibility of applying:

- Unsealing of pit-banks of return shafts ('Zofia' and 'Zbyszek') at the phase, when their main fans are operating but at the same time stoppings with windows are already built (within works preceding further liquidation works, including shafts filling),
- Partial or complete unsealing of regulating doors situated in individual ventilation sub-systems (this applies also to chambers in the area of central shafts).

Because of endogenous fires hazard, it is recommended to analyse the regulating doors location and the tightness of available insulation stoppings, considering changes of potential drop distribution, in particular, in the 'Wschodni' shaft sub-system.

Parallel to carried out analyses of current ventilation conditions and possible corrections of previous decisions, the recognition and control of gas conditions are a very significant issue. The recognition and decisions are required mainly in the case of migration of goaf gases and those filling dammed complexes of workings and goafs, creating specific hazard for the mine operation and staff.

5 Conclusion

An example of a staged abandonment of a complex system of an underground mine workings has been presented. The case study uses a mine ventilation system simulator Ventgraph to make a forecast of the ventilation conditions and assess the effect of planned actions. The plan uses both a general frame and knowledge on specific conditions of the mine. It has been implemented in practice and met the assumed requirements.

Acknowledgements Research presented in this paper was supported by the MERIDA EU RFCS RFCR-CT-2015-00004 and Polish MNiSW W52/FBWiS/2016 research project and Polish MNiSW W52/FBWiS/2016.

References

Bystroń, H.: An approach to mine ventilation based on the aerodynamic potential of ventilating air treated as a mixture of dry air, water vapour and liquid water droplets. In: Proceedings of the International Mine Ventilation Congress, Chapter 1. EMAG, Cracow, Katowice, Poland, 17–22 June 2001. pp. 1–8

Dziurzyński, W., Pałka, T., Krawczyk, J.: Ventgraph for Windows—Ventilation Engineer's Program System for Analyzing the Ventilation Network Under Normal and Emergency Conditions—Simulation of Transient Flow of Air and Fire Gases—Manuals, pp. 1–100. Transaction of Strata Mechanics Research Institute, Kraków, Poland (2013)

Dziurzyński, W., Pałka, T., Krach, A.: A reliable method of completing and compensating the results of measurements of flow parameters in a network of headings. Arch. Min. Sci. **60**, 3–24 (2015)

Dziurzyński, W., Krach, A., Pałka, T.: Airflow sensitivity assessment based on underground mine ventilation systems modeling. Energies **10**, 2–15 (2017)

Pritchard, C.J.: Validation of the ventgraph program for use in metal/non-metal mines. In: Proceedings of the 13th United States/North American Mine Ventilation Symposium, Sudbury, Canada, June 2010. pp. 455–462

Part VII
Emerging Monitoring and Measurement Technologies

Data Acquisition System for Position Tracking and Human-Selected Physiological and Environmental Parameters in Underground Mine

P. Stefaniak, J. Wodecki, A. Michalak, A. Wyłomańska and R. Zimroz

1 Introduction

Rapid development of communication and electronics systems enables to develop technology for tracking human physical position and activity together with the change of physiological and environmental parameters. The harsh conditions of underground mine relate mainly to high temperature, humidity, dust, salinity, and concentration of various gases in the air (Bascompta and Sanmiquel 2015; Hartman et al. 2012). There exist many areas where levels of toxic gases (e.g., H_2S and CO) might locally exceed limit values. Furthermore, it needs to be pointed out that miners undergo intensified stress due to the presence of various factors in underground mine (Ryan et al. 2017). Such working conditions in combination with medium to high energy levels required in mining working conditions may induce severe organism functioning disorders and limited physical fitness and motor coordination of workers (Dosinas et al. 2006), apart from increased body temperature and heart rate.

Similar technologies exist (Dosinas et al. 2006; Shaffer et al. 1992; Hol et al. 2007; Sabatini 2011; Aggarwal and Ryoo 2011; Paiyarom et al. 2009;

P. Stefaniak · J. Wodecki · A. Michalak · A. Wyłomańska (✉) · R. Zimroz
KGHM Cuprum Ltd., Research and Development Centre, Lubin, Poland
e-mail: awylomanska@cuprum.wroc.pl

P. Stefaniak
e-mail: pkstefaniak@cuprum.wroc.pl

J. Wodecki
e-mail: jwodecki@cuprum.wroc.pl

A. Michalak
e-mail: amichalak@cuprum.wroc.pl

R. Zimroz
e-mail: rzimroz@cuprum.wroc.pl

E. Widzyk-Capehart et al. (eds.), *Proceedings of the 18th Symposium on Environmental Issues and Waste Management in Energy and Mineral Production*,
https://doi.org/10.1007/978-3-319-99903-6_21

Pantelopoulos and Bourbakis 2010); however, the majority of them are applied in sport and healthcare sector and are not easily applicable to underground mine reality. Those solutions are designed to operate in isolated, controlled conditions, and the influence of miner physical activity on its physiological parameters in harsh and varying environmental conditions has not been recognized in detail yet. Currently, applied methods of tracking miners are based on RFID technology, which provides only data to identify miner's presence in certain zone, without any positioning in three-dimensional space. In addition, application of inertial sensor in the proposed system allows to recognize the type of performing physical activity, which is not possible using simple zone localization. The proposed system can enable measuring location of the miner, basic physiological parameters (temperature, pulse), and environmental parameters (temperature, humidity, gas presence).

It is also relevant to consider gas concentration analysis of mining air. Fusion of such data with location information allows to perform context analysis, which may deliver many information necessary to determine as yet unknown relationships between behaviors of an organism corresponding to different physical activities in underground mine. Ongoing monitoring of air parameters allows to estimate real parameters of local atmosphere, regularities between these parameters, and employee's tiredness or type of work.

Since wireless telecommunication in underground corridors is almost impossible to maintain for technological reasons (even wired communication is limited to emergency landlines), it is impractical for now to deploy online monitoring system that transmits data from the sensors in real time. To address this difficulty, offline monitoring system has been proposed.

This paper describes the development of a prototype monitoring system to measure physical activity of miners, their health parameters, and selected environmental parameters of their workplace including post-factum location in corridor. It presents initial results of testing of measuring platform in underground mine to validate a technical functionality of measurement layer with the focus on the motion tracking.

2 Main Requirements for the System

Requirements for the system assume technological simplicity but are mostly driven by the need of measuring certain physical parameters of the environment and the workers themselves (see Fig. 1).

One of the main objectives of the system is registering motion data that enable path reconstruction possible to be projected on a map. In order to perform such measurements, there is a need to employ so-called inertial measurement units (IMUs) that consist of accelerometers, gyroscopes, and magnetometers. Fusing raw data from those sensors using appropriate mathematical concepts allows to obtain displacement information, which is necessary for path reconstruction. Moreover, in

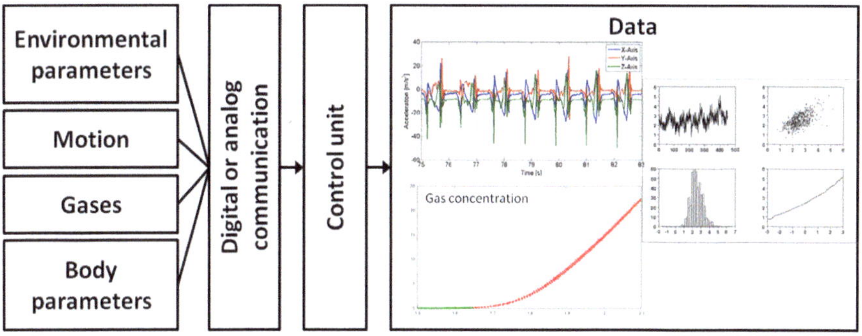

Fig. 1 Functional concept of the system

Table 1 Measured parameters

Group	Parameter
Environmental parameters	Air temperature Air pressure Air humidity
Gases	H_2S concentration CO concentration
Physiological parameters	Body temperature Pulse
Motion parameters	IMU on the head IMU on the foot

order to record motion data additional data needs to be collected as listed in Table 1.

3 System Design

3.1 Requirements and Measurement Parameters

Taking into consideration the requirements for the system, commercially available components have been selected to build the prototype (see Fig. 2; Table 2).

3.2 System Architecture

Elements have been assembled according to the architecture shown in Fig. 3. Sensors measuring body parameters and concentrations of gases have been connected to Arduino board via analog bus, which reads voltage levels provided by the

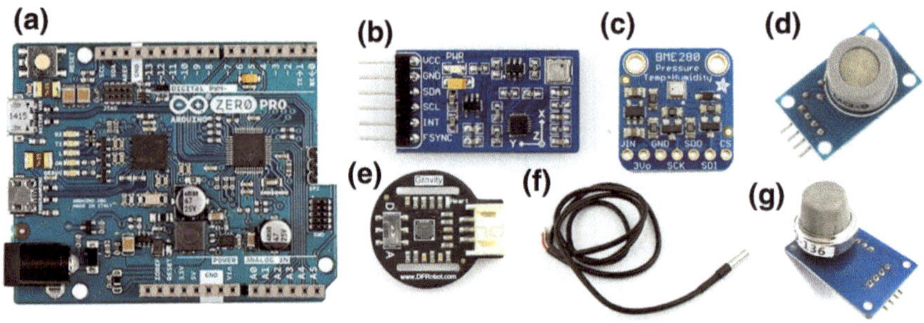

Fig. 2 Functional components of designed system: **a** control unit (Arduino platform), **b** inertial sensor, **c** temperature, pressure, and humidity sensor, **d** CO sensor, **e** pulse sensor, **f** body temperature sensor, and **g** H_2S sensor

Table 2 System components

Component	Role
(a) Arduino M0 Pro	Control platform
(b) Waveshare IMU 10DoF	Inertial measurements
(c) Adafruit BME280	Air temperature Air pressure Air humidity
(d) MQ-9 sensor	CO concentration
(e) DFRobot Gravity	Pulse
(f) DS18B20 sensor	Body temperature
(g) MQ136 sensor	H_2S concentration

sensors. Remaining sensors communicate with the processor via digital buses: I^2C and SPI that operate using their own standardized protocols.

3.3 *Data Acquisition and Processing*

Sample rates of individual parameters are also very important to consider. Since environmental and vital parameters are by definition slowly varying processes, it is not necessary to perform such measurements with high speeds. Hence, for those parameters, sample rate is set to 1 Hz. The only exception is the pulse. For the reasons of memory management, pulse sensor operates in digital mode. It means that instead of transmitting complete data of the pulse wave (which requires appropriate processing afterward to identify the pulse rate), sensor processes the data internally and only transmits a single data packet at each heartbeat. This method simplifies system integration and significantly reduces the amount of data to be transmitted and stored. Hence, pulse parameter has no constant sample rate but is managed asynchronously with the mechanism known as "interrupts" that triggers the sample transmission as soon as the heartbeat is detected. In case of motion

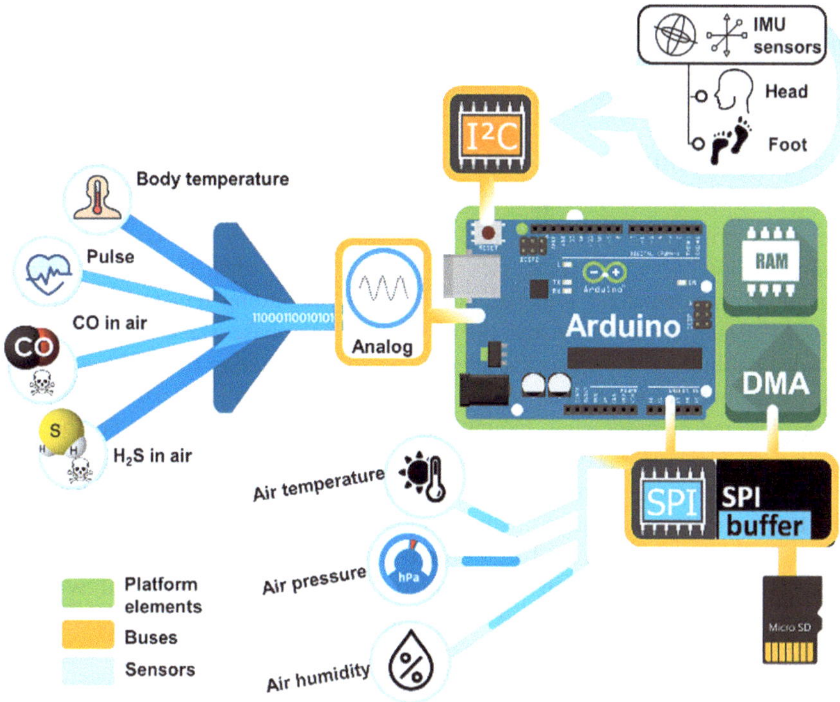

Fig. 3 System architecture

measurements, it is crucial to capture every slight movement of the sensor to be able to correctly reproduce motion trajectory. If any part of data is missing, motion trajectory will be incorrect. For example, if during the measurement, data acquisition fails when a person is making a turn while walking, data from before turning will be followed by data from after turning.

Hence, from the data point of view, the turn never occurred and person was walking in straight line all the time. In such case, the events after the erroneous measurement cannot be considered valid. Even a loss of data from very small movement will cause an avalanche effect. Since such measurements are based on vibrations and rotations (which are varying very fast), it is crucial to sample the data with sufficiently high speeds. For practical reasons, sample rate of IMU sensor was set to 100 Hz.

Another aspect of motion tracking is proper data interpretation. Since magnetic measurements are inapplicable in underground conditions, it is important to be able to operate on accelerometric and gyroscopic data only. However, values measured directly by sensors cannot be used for analysis, but have to "work together"; hence, there is a need to apply data fusion algorithm to appropriately convert raw data to interpretable equivalent of accelerometer data or linear accelerations and Euler angles (interpretable equivalent of gyroscopic data). In such operation, appropriate data smoothing is essential (Wodecki et al. 2018). This task is performed by employing Kalman-based Altitude and Heading Reference System (AHRS) algorithm (Marmion 2006).

4 Functionality Testing of Inertial Measurements

The initial prototype testing was performed in a laboratory in a controlled environment. Slowly varying processes (e.g., gases and other environmental and physiological parameters) imposed no difficulties regarding registering and validation. For inertial measurements, the detailed testing phase was conducted, which included calibrating the sensors, establishing optimal filtering methods, stationary states elimination, and other signal-processing-related aspects of proper data acquisition and preprocessing. Example of registered inertial data can be observed in Fig. 4.

The system was validated during a measurement, and the aim of this phase was to determine if it is possible to obtain data that could be correlated with the planned route. Test route was selected based on the layout of underground corridors in the mine. Route was mapped in the laboratory, and a person equipped with the acquisition system walked along it taking a measurement. Samples were registered with the sampling frequency of 500 Hz, which translates into about 13,000 samples per pass that took the person about 25 s. Every individual measurement session included a single walk along the route (there is no averaging of multiple passes). Collected data was superimposed on the map of planned route to verify if it is possible to identify the shape of planned route. Result of the route comparison with the measurement is presented in Fig. 5. Green arrows in Fig. 5 correspond to red ones in Fig. 6.

From Figs. 5 and 6, it can be seen that registered route corresponds almost perfectly to the planned one. Some level of roundness in registered route comes

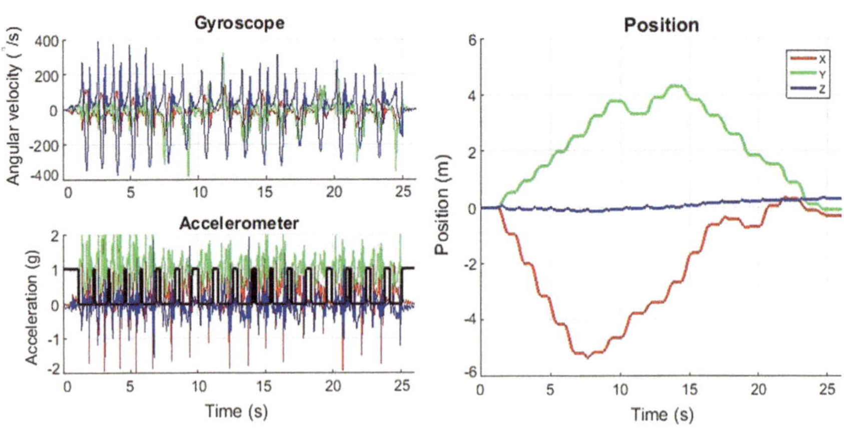

Fig. 4 Exemplary inertial signals

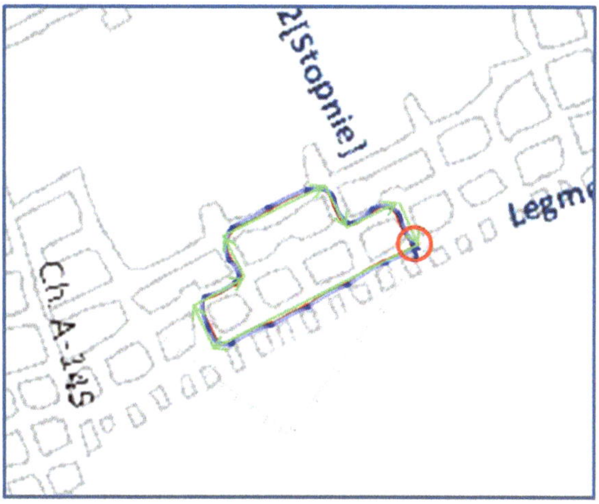

Fig. 5 Route correlation experiment. Green arrows denote segments of planned route, blue data is a set of versors for the measured points, and red circle is the point of start and finish

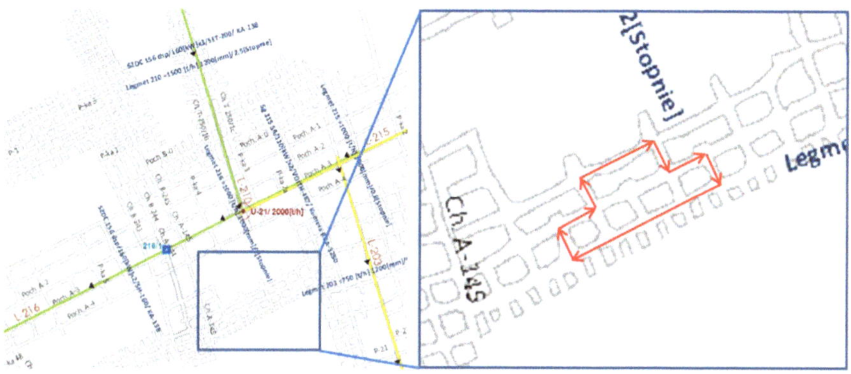

Fig. 6 Example of desired data visualization

from the internal drift of a gyroscope. The drift is already reduced by a large amount by calibrating the sensor as well as possible and introducing selective drift elimination in the post-processing algorithm, but in reality it is practically impossible to eliminate it completely.

5 Conclusions

In this paper, offline monitoring system has been proposed. It registers basic environmental parameters, miner's vital functions, and their relative position in three-dimensional space. Analyzing such data can allow to obtain better information about miners' location, e.g., for the purpose of coordinating search and rescue actions. Another application is the analysis of worker activity, fatigue estimation, difficult areas identification, etc. System has been tested in laboratory conditions with satisfying results. Further work includes performing tests in real-life mining conditions underground, as well as further software-wise system optimization.

Acknowledgements This work is supported by the KGHM Cuprum R&D Ltd. statutory grant "Analiza możliwości technologicznych nawigowania i analizy parametrów aktywności fizycznej pracowników dołowych w trudnych warunkach środowiskowych kopalni podziemnej".

References

Aggarwal, J., Ryoo, M.: Human activity analysis: a review. ACM Comput. Surv. **43**, 16, 1–16, 43 (2011). https://doi.org/10.1145/1922649.1922653

Bascompta, M., Sanmiquel, L.: Determination of the environmental conditions in an underground mine. In: Proceedings of the 2nd International Conference on Mining, Material and Metallurgical Engineering (MCM 2015) (2015)

Dosinas, A., Vaitkunas, M., Daunoras, J.: Measurement of human physiological parameters in the systems of active clothing and wearable technologies. Elektronika ir Elektrotechnika **71**, 77–82 (2006)

Hartman, H.L., Mutmansky, J.M., Ramani, R.V., Wang, Y.: Mine Ventilation and Air Conditioning. Wiley (2012)

Hol, J.D., Schön, T.B., Luinge, H., Slycke, P.J., Gustafsson, F.: Robust real- time tracking by fusing measurements from inertial and vision sensors. J. Real-Time Image Proc. **2**, 149–160 (2007)

Marmion, M.: Airborne Attitude Estimation Using a Kalman Filter, pp. 1–85. The University Centre of Svalbard, Longyearbyen, Norway (2006)

Paiyarom, S., Tungamchit, P., Keinprasit, R., Kayasith, P.: Activity monitoring system using dynamic time warping for the elderly and disabled people. In: 2nd International Conference on IEEE Computer, Control and Communication, 2009, pp 1–4. IC4 2009

Pantelopoulos, A., Bourbakis, N.G.: A survey on wearable sensor-based systems for health monitoring and prognosis. IEEE Trans. Syst. Man Cybern. Part C (Appl. Rev.) **40**, 1–12 (2010)

Ryan, A., et al.: Heat stress management in underground mines. Int. J. Min. Sci. Technol. **27**, 651–655 (2017)

Sabatini, A.M.: Estimating three-dimensional orientation of human body parts by inertial/magnetic sensing. Sensors **11**, 1489–1525 (2011)

Shaffer, G.K., Stentz, A., Whittaker, W.L., Fitzpatrick, K.W.: Position estimator for underground mine equipment. IEEE Trans. Ind. Appl. **28**, 1131–1140 (1992)

Wodecki, J., Michalak, A., Stefaniak, P.: Review of smoothing methods for enhancement of noisy data from heavy-duty lhd mining machines. E3S Web Conf. **29**, 00011 (2018) (EDP Sciences)

Development of a Dust Violation Control Tool from Plant Data

Mustafa Erkayaoglu

1 Introduction

Mining is an equipment-intensive and interdisciplinary industry that incorporates different engineering disciplines. Mineral processing is an essential stage of mining where the material produced by drilling and blasting for surface operations is enriched and waste material is separated as much as possible. Mineral processing plants have various equipment systems used for size reduction, materials handling, and separation processes. In surface mining, blasted material is transported by haul trucks and conventionally handled by belt conveyors throughout different stages of mineral processing. Belt conveyors and most of the machinery in mineral processing plants are controlled and monitored by sensors and systems. This enables the close control of equipment both by means of production and maintenance.

Another major reason of utilizing sensor systems in the mining industry is maintaining the compliance to environmental standards. Sensors that are reading different environmental parameters are commonly visualized on supervisory control and data acquisition (SCADA) systems where value readings are continuously updated and utilized as production information in the plant (Smit 2002). This enables the engineer in the control room to take action in case an alarm is defined for predefined cases. However, the amount and frequency of data available in modern mines reached a level that human controllers cannot cope with. Therefore, smart systems and integrated data have to be part of daily mine management as much as possible. This study introduces an implementation of sensor data in mineral processing plants where the environmental monitoring system and the historian were integrated to control the operation of the water sprays on the belt conveyors.

M. Erkayaoglu (✉)
Mining Engineering Department, Middle East Technical University, Ankara, Turkey
e-mail: emustafa@metu.edu.tr

E. Widzyk-Capehart et al. (eds.), *Proceedings of the 18th Symposium on Environmental Issues and Waste Management in Energy and Mineral Production*,
https://doi.org/10.1007/978-3-319-99903-6_22

2 Methodology

Continuous development of the technology related to equipment and IT systems has enabled modern mining operations to utilize data as part of their decision-making process. However, this requires a reliable and integrated layer of data that is developed and maintained according to management requirements. The methodology of this study is based on data integration of different systems to monitor the functionality of spray systems that are used to suppress the dust on belt conveyors. Prior to data integration, available data sources were identified and characterized to determine the potential integration points by means of location, time, or other data dimensions. There are different types of data available in the mining industry, such as relational and process type that can be integrated for Business Intelligence (BI) purposes.

2.1 Process Type of Data

Process or industrial data are the collection of sensor or control data from real-time industrial processes, often called embedded systems. Programmable logic controllers (PLCs) are computers that control and monitor industrial systems with the help of human 'operators' who monitor and tweak the industrial processes through SCADA systems (Alphonsus and Abdullah 2016). Both the control interactions and the conditions throughout the industrial process can be recorded and stored either through regular sampling into a relational database or through proprietary industrial data storage tools. The signal measures are recorded as 'tags,' where a tag's measure value, such as the amperage from a particular conveyor motor is recorded once per minute. This process view of data is particularly suitable for real-time trending or finding relationships between different quantitative analog variables. Significant innovation is emerging from the increasingly more accessible process data. Process data can be regarded as simplistic tagging data, where much of the meaning of the measure relates to the metadata of the tag itself necessitating a detailed description of the tag/measure and its many implications within the process; hence, process data can only be considered partially structured or 'semi-structured.'

This primary type of data for the mining industry has a completely different structure and characteristics than relational data sources. Process control systems, environmental monitoring systems, and SCADA systems are the main sources of process data that are present in mineral processing plants. Crushers, belt conveyors, mills, and other critical components of the value stream are equipped with sensors or tags. These tags collect data in a more primitive way than relational data sources. However, the data frequency is much higher which requires a different IT strategy to process the data.

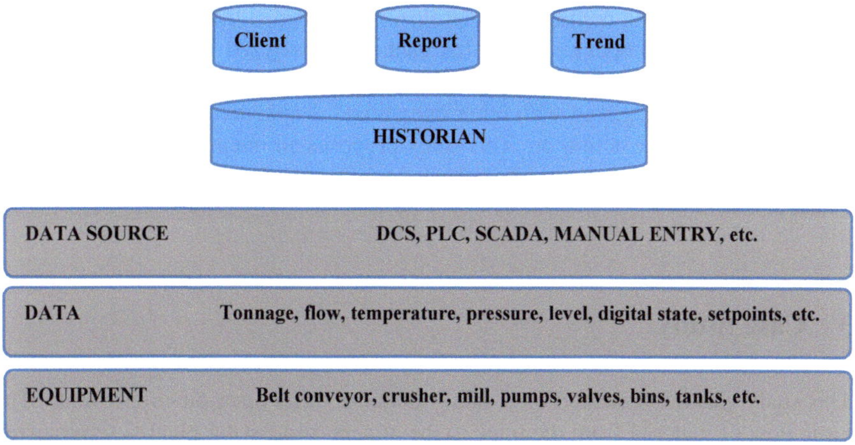

Fig. 1 Working principle of historians

Mineral processing-related data typically consist of a measurement and a time stamp. Power consumption, tons per hour, and run time are basic measures that are tracked. As the frequency of process data is much higher than that of relational data, process data from sources are handled by hardware components called data historians. The basic working principle of historians is illustrated in Fig. 1.

Historians are capable of rapidly reading and writing data records. Process-related data is small in size, but its frequency creates a vast amount of data. This is one of the main challenges in utilizing process type of data in daily mine management. Isolating the data to certain departments limits the potential for creating value for the whole value chain. Therefore, common performance metrics between different departments should be defined and visualized in a user-friendly environment. Historians are also capable of enabling advanced analytics by integrating into corporate data management systems (Chardin et al. 2012).

2.2 *Balanced Scorecards*

Balanced scorecards are BI tools that provide an overall summary of the performance of a system. Integrated data from different data sources can be evaluated on performance metrics, also known as key performance indicators (KPIs) (Kaplan and Norton 1996). There are different ways of defining performance metrics based on the goals defined; however, it is of major importance to utilize BI tools as cultural enablers to change management. Although the balanced scorecard approach is agreed to beneficial for management, its positive impact depends on the way it is implemented in management (Papalexandris et al. 2005). The impact of utilizing data and keeping it in the center of daily mine management can easily be tracked.

This also supports the continuous improvement strategies that are implemented in corporate mining companies. Data that cannot be analyzed by a human operator cannot be considered as a reliable tool for decision making. Therefore, process type of data has to be integrated with other operational data sources and summarized to make it easier to take action on. This study proposes an integration of process type of data that can be used as part of a balanced scorecard of a mineral processing plant manager.

3 Case Study

This study was implemented on a data warehouse of an open-pit copper mine. The basic flow of material from the mine to the mineral processing plant is summarized in Fig. 2.

The mineral processing plant that provided the data for this study had issues related to their environmental compliance. The main belt conveyors that were transporting the coarse ore, the fine ore, and the mill feed were stated to be generating excessive an amount of dust although they were equipped with water sprays. The head and tail dust suppressors were configured to be initiated when a threshold dust concentration was observed. However, the system was not triggered at certain times although critical dust concentrations were recorded. In order to provide operational data to the engineers, the tags of the belt conveyors and the environmental management system were integrated. Integration of data for a specific purpose is one of the key components of continuous improvement in the mining industry (Bascur and Soudek 2010). As the data warehouse used in this study was

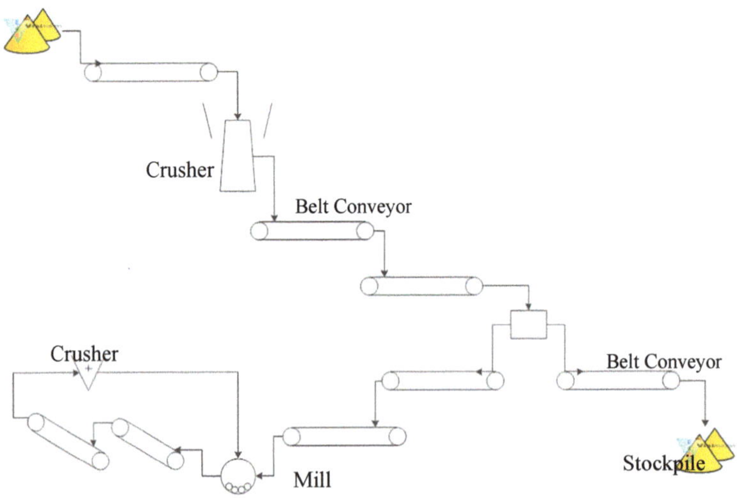

Fig. 2 General overview of mineral processing plant

real-time, the information whether the dust suppression system was active or not could be seen instantaneously. A historical trend of the integrated data is summarized for a 24 h basis in Fig. 3.

As it can be seen, the tons per hour rating of the belt conveyors, the flow of head and tail sprays of the belt conveyors, and the operation of the dust collector were

	Spray Bars						
	CV3		CV4		CV5		Totals
Total Exceptions	9		0		0		9
	Head	Tail	Head	Tail	Head	Tail	Avg Compliance (%) and Hours
Compliance (%)	25.0%	62.5%	95.8%	100.0%	100.0 %	91.7%	79.2%
Hours Not Spraying w/ Feed Running	18	9	1	0	1	2	5.17

	Bag House Pressure Differential While Running
Exceptions	6
Max Differential	2.10
Min Differential	1.41

Hours	Max Tag Value	Min Tag Value	Avg cv03 TPH	Avg CV03 Head Spray Flow	Avg CV03 Tail Spray Flow	Avg cv04 TPH	Avg CV04 Head Spray Flow	Avg CV04 Tail Spray Flow	Avg cv05 TPH	Avg CV05 Head Spray Flow	Avg CV05 Tail Spray Flow	Avg Dust Collector Diff Pressure
00:00	1.70	1.70	1660.78	0.00	7.79	1493.08	25.19	3.69	0.00	2.27	1.86	1.70
01:00	1.44	1.44	1899.90	0.00	9.86	1713.82	24.75	3.77	0.00	5.49	0.00	1.44
02:00	1.79	1.79	2981.71	0.00	5.40	1944.24	23.25	4.01	3093.65	3.36	3.89	1.79
03:00	1.49	1.49	2971.66	0.00	0.00	1537.91	30.75	3.99	3024.02	3.69	4.38	1.49
04:00	1.55	1.55	2777.25	0.00	5.24	2343.41	22.76	3.93	2889.74	2.42	1.06	1.55
05:00	1.83	1.83	3173.74	0.00	9.55	3085.27	25.35	3.76	0.00	0.00	0.00	1.83
06:00	1.58	1.58	0.00	0.00	4.87	0.00	22.94	3.77	0.00	0.00	0.00	1.58
07:00	1.80	1.80	1093.27	0.00	0.00	0.00	24.75	3.76	0.00	0.00	0.00	1.80
08:00	1.65	1.65	1782.35	0.00	0.00	1634.52	23.25	4.10	0.00	0.00	0.00	1.65
09:00			3123.68	0.00	0.00	2329.33	21.76	4.08	3114.51	8.71	1.38	
10:00			3163.62	0.00	0.00	1171.09	24.55	4.24	3255.99	2.36	2.72	
11:00	2.01	1.81	3280.62	0.00	0.00	3241.96	22.94	4.65	0.00	0.00	0.00	1.91
12:00	1.58	1.55	3178.66	0.00	8.36	2597.65	38.10	4.57	3169.25	6.01	1.06	1.57
13:00			3193.68	0.00	9.12	0.00	29.72	2.29	3262.64	3.27	0.00	
14:00			3292.09	0.00	8.91	3258.15	0.00	4.80	0.00	0.00	0.00	
15:00			2299.44	4.28	3.09	2154.61	24.06	4.35	0.00	0.00	0.00	
16:00	1.55	1.55	0.00	0.00	0.00	0.00	0.00	0.00	0.00	0.00	0.00	1.55
17:00			0.00	0.00	0.00	0.00	0.00	0.00	0.00	0.00	0.00	
18:00	1.17	0.00	0.00	0.00	0.00	0.00	3.54	0.00	0.00	0.00	0.00	0.75
19:00			0.00	0.00	0.00	0.00	24.37	4.12	0.00	0.00	0.00	
20:00	2.10	1.52	2826.88	0.00	0.00	2699.72	21.74	4.57	2321.31	5.26	0.00	1.87
21:00	1.43	1.43	2908.42	0.00	0.00	0.00	0.00	1.85	2899.09	5.12	5.29	1.43
22:00			2575.52	0.00	0.00	1946.41	27.12	3.93	1337.72	4.06	2.83	
23:00	1.90	1.41	2981.30	0.00	1.21	0.00	33.56	3.52	3002.11	3.24	4.92	1.65

Fig. 3 Historical trend of dust suppression system

Table 1 Monthly summary of dust suppression system activity

	# of violations	% of time (%)
CV-03 head spray	119	35
CV-03 tail spray	9	3
CV-04 head spray	0	0
CV-04 tail spray	8	2
CV-05 head spray	1	0
CV-05 tail spray	126	38
Dust collector	8	3

integrated on a common basis of time. This enabled the engineers to monitor the operation of the dust collectors and water sprays based on the current production rate. It was observed that the exceptions of the water spray systems working were critical, especially crucial at the conveyor that handled coarse ore. As the common dimension to integrate these data sources was time, the level of granularity could be changed from an hourly basis to a minute or aggregated to a month as seen in Table 1.

As a result, the developed tool for integration of the belt conveyors, water sprays, and the dust collector was implemented into the scorecards of the processing plant manager. This led to a decrease of the malfunction of the water sprays in a considerably short time. Enabling engineers to utilize integrated data resulted in a process improvement, and this tool started to be used as part of the environmental management system on site.

4 Conclusion

This study presented a case study of data integration at a mineral processing plant focusing on the dust suppression systems installed on belt conveyors. The belt conveyor responsible for handling fine and coarse ore was equipped with water sprays at their head and tail sections. An excess amount of dust in the mineral processing plant was observed, and efficient utilization of the water sprays was questioned. The mineral processing plant failed to comply with environmental standards and was given notices by legal authorities. Data from the plant control system and the environmental monitoring system were integrated to control whether the water sprays were activated. Although plant type of data is commonly visualized as a time series in control rooms, making decisions or taking actions on this real-time data is challenging. The integration of this data with other related data sources enabled the engineers to question the functionality of available technology in the mineral processing plant. Visualizing and creating a measure from this data led to a continuous improvement action once the balanced scorecard of the mineral processing plant manager included this information. This proves the necessity of data integration for cases that are monitored by high-frequency and real-time data.

Human operators in the control room cannot be expected to make correct decisions by trying to analyze and evaluate multiple sources of data. The nature of plant type of data in the mining industry exposes the lack of available manpower to use this information in daily management and decision making. The tool developed within the scope of this study can be considered as an example of the potential benefit of data integration in environmental management of mineral processing plants. Other available data sources can be integrated to create a flexible environment to enable engineers to utilize available technology more efficiently.

References

Alphonsus, E.R., Abdullah, M.O.: A review on the applications of programmable logic controllers (PLCs). Renew. Sustain. Energy Rev. **60**, 1185–1205 (2016)

Bascur, O., Soudek, A.: Real-time integration of mining and metallurgical information for efficient use of energy and water. In: XXV International Mineral Processing Congress (IMPC), pp. 3123–3131. Brisbane, Australia (2010)

Chardin, B., Lacombe, J.-M., Petit, J.-M.: Data historians in the data management landscape. In: Technology Conference on Performance Evaluation and Benchmarking, pp. 124–139 (2012)

Kaplan, R.S., Norton, D.P.: Translating Strategy into Action—The Balanced Scorecard. Harvard Business School Press (1996)

Papalexandris, A., Ioannou, G., Prastacos, G., Soderquist, K.E.: An integrated methodology for putting the balanced scorecard into action. Eur. Manag. J. **23**(2), 214–227 (2005)

Smit, A.: Development of an integrated asset management system for a metallurgical plant. In: Value Tracking Symposium, pp. 157–165. Brisbane, Australia (2002)

Finding the Right Time to Mine: A Real Options Analysis of Landfill Mining Projects

M. Menegaki, D. Damigos, A. Benardos and D. Kaliampakos

1 Introduction

The depletion of the mineral resources, together with the environmental footprint of former and operating mining sites, puts a lot of stress in mining companies. Technological evolution may allow new types of exploitation in the near future (i.e., extraterrestrial mining), or re-exploitation of low-grade resources. Nevertheless, in terms of sustainability, the sector should look into the opportunities that appear today in order to continue not only to be competitive but also to respond to the increasing social requirements. Landfill mining could be such an opportunity under specific circumstances (Menegaki et al. 2015).

In general, landfill mining is described as "a process for extracting minerals or other solid natural resources from waste materials that previously have been disposed of by burying them in the ground" (Krook et al. 2012). Technically, landfill mining (LFM) is the process of excavating from operating or closed solid waste landfills by employing the method of open cast mining and sorting the unearthed materials for recycling, processing, or for other dispositions (Lee and Jones 1990; Cossu et al. 1996; Hogland et al. 1997; Carius et al. 1999).

There are several studies providing data with respect to the quantities of materials disposed of in landfills. Typically, besides organic matter, municipal solid waste landfills contain significant amounts of paper, plastic, wood, rubber, glass, leather and other textiles, and earth construction materials and metals (Obermeier et al. 1997; Hogland et al. 2004; Kurian et al. 2007). For example, according to Kapur and Graedel (2006), the quantity of copper located in landfills and other waste repositories around the world is about 300 million tonnes, an amount corresponding to more than 30% of the remaining reserves in known ores (Frändegård

M. Menegaki (✉) · D. Damigos · A. Benardos · D. Kaliampakos
School of Mining and Metallurgical Engineering, National Technical University of Athens, Athens, Greece
e-mail: menegaki@metal.ntua.gr

E. Widzyk-Capehart et al. (eds.), *Proceedings of the 18th Symposium on Environmental Issues and Waste Management in Energy and Mineral Production*,
https://doi.org/10.1007/978-3-319-99903-6_23

et al. 2013). These waste deposits could be the source of a significant stream of secondary materials and energy (Jones et al. 2013) that could comprise the "ore" of the future. Nevertheless, like any other economic activity, LFM has to be economically feasible, and, so far, most of the findings are contradictory. The main reason is that the economic factors associated with LFM are numerous and complex. To this end, this paper aims at estimating the economic feasibility of LFM projects by means of a modern management approach, namely the Real Option Valuation (ROV). For the purposes of the analysis, a hypothetical "typical" landfill site is examined and different alternatives are formed.

2 Methodology and Case Study Background

2.1 *Conventional Valuation Tools*

The discounted cash flow (DCF) analysis is probably the most widely used technique of investment analysis, in general. The DCF method values a proposed project based on a number of performance criteria, the most commonly applied of which are the Net Present Value (NPV) and the Internal Rate of Return (IRR).

Regardless of the criterion used, the NPV and IRR values of any project (and consequently the investment decision) are basically influenced by two factors, namely the timing of project cash flows and the uncertainty of receiving the expected cash flows (Samis et al. 2006; Poulimenou and Damigos 2015). In order to tackle the uncertainty involved in the estimates, the performance criteria are explored by means of sensitivity and probabilistic risk analysis (Menegaki et al. 2017). Sensitivity analysis is carried out by varying one variable at a time and allows determining the effect of each variable on the NPV and IRR indices. The probabilistic analysis involves assigning a probability distribution to each of the critical variables of the DCF model based on literature data, experimental data, expert opinions, etc., and conducting simulations by means of the Monte Carlo approach (Damigos et al. 2016). The calculations for all combinations of sampled values are then used to develop probability distributions of the NPV and IRR indices offering more comprehensive information about the risk profile of the project.

2.2 *Real Options Valuation*

Given the difficulties and uncertainties involved, the conventional investment analysis tools, i.e., the DCF approach, cannot value embedded risks and opportunities correctly because they operate within a "now-or-never" decision-making framework. As de Neufville (de Neufville 2003) mentions: *"...from the perspective of normal engineering practice, the new, options-based approach is fundamentally - perhaps even cataclysmically - different. Engineers are trained to reduce risk, to prevent failures…", while "…the framework of options thinking, however, recognizes that uncertainty adds value to options, and can be viewed as a positive element…".* To this direction, over the last three decades, modern management approaches, such as, the Real Option Valuation (ROV), begin to receive more and more attention in order to minimize the different kinds of risks that threaten irreversible investments.

ROV has its roots in the theories developed by Black and Scholes (1973) and Merton (1973). In general, a real option is a right, not an obligation, to take an action on an underlying asset. In order to acquire the option, an option price or premium has to be paid. With regard to investment valuation, this may involve abandoning or expanding a project, or deferring the decision until the uncertainty becomes resolved (Kodukula and Papudesu 2006; Mun 2006). In general, the higher the uncertainty, the higher is the expected option value by the model (Kodukula and Papudesu 2006). To date, a number of ROV applications in mining have been published (Samis et al. 2006; Brennan and Schwartz 1985; Damigos and Menegaki 2011; Dimitrakopoulos and Sabour 2007; Kelly 1998; Moyen et al. 1996; Sabour et al. 2008; Slade 2001).

The solution of the ROV model is possible through multiple techniques, the selection of which depends on the availability of input data, the type of the option (e.g., American or European), the desired simplicity, etc., as detailed by Kodukula and Papudesu (2006) and Mun (2006). In the present study, the analysis is based on Monte Carlo path-dependent simulations and binomial lattices. The calculation of ROV starts with the computation of the "current underlying asset value" by means of traditional discounted cash flow approaches where the investment cost reflects the "strike price." The option value lattice is then estimated using the uncertainty involved and the decision flexibility. Starting with the present value of expected future cash flows at year 0 (*S*0), the lattice of the underlying asset is created by multiplying *S*0 with the up (*u*) and down (*d*) factors, as illustrated in Fig. 1.

The basic inputs required are as follows (Mun 2006): the present value of the underlying asset at year 0 (*S*0); the present value of the implementation cost of the option (*X*); the volatility of the natural logarithm of the underlying cash flow returns (σ, in percentage); the time to expiration (*T*, in years); the risk-free rate of return (*rf*, in percentage); and the stepping time, i.e., timescale between steps (δt). The calculation of the up (*u*) and down (*d*) factors and the risk-neutral probability measure (*p*) is accomplished according to Eqs. 1, 2, and 3:

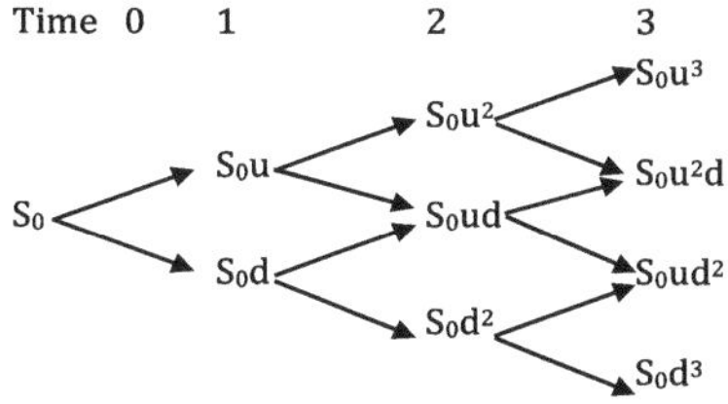

Fig. 1 Binomial lattice of the underlying asset value

$$u = e^{\sigma\sqrt{\delta t}} \tag{1}$$

$$d = e^{-\sigma\sqrt{\delta t}} \tag{2}$$

$$p = \frac{e^{(rf*\delta t)} - d}{u - d} \tag{3}$$

Further, the volatility factor, σ, needs to be estimated. There are several methods for this direction. In this case, the logarithmic present value approach was employed (Mun 2006).

The calculation of the option valuation lattice requires a two-step procedure, which is called backward induction. The calculation starts with the terminal nodes. Given that we wish to maximize the expected benefits, if the execution of the option (i.e., $S - X$) yields profit, then the value becomes equal to that profit. If the cost exceeds the benefit of execution, then the value becomes equal to 0. The second step involves the calculation of the intermediate nodes, using the P value, according to Eq. 4:

$$\text{node value} = [(p)\text{up} + (1 - p)\text{down}]e^{(-rf*\delta t)} \tag{4}$$

Hence, the intermediate values become equal to the discounted weighted average of future option values, and the decision whether to keep the option open or exercise the option is defined by the maximum of the two outcomes, namely the execution of the option (i.e., $S - X$) and the value of the intermediate node.

2.3 Technical and Financial Assumptions

The DCF analysis of an LFM project usually takes into consideration the following factors (Menegaki et al. 2017; Damigos et al. 2016; Danthurebandara et al. 2015; Frändegård et al. 2015; Zhou et al. 2015):

- capital costs, e.g., inventory and permit costs, site preparation, purchase of equipment;
- operating costs, such as rental of equipment (if rented), labor costs, energy costs, maintenance costs; and
- revenues from: recyclable and reusable materials, recovered airspace (in case that landfill continues to operate), avoided costs of post-closure care (in case of full site reclamation), etc.

In this study, benefits from energy recovery, redevelopment of the landfill area, and reduction in the costs of landfill closure and post-closure care and monitoring are not considered, owing to the existing conditions in Greece (e.g., RDF energy utilization in Greece is not possible) and the technical assumptions used.

The basic technical and financial assumptions, which are briefly discussed hereinafter, are based on a previous research by the authors (Menegaki et al. 2017). The analysis considers a typical urban landfill, with a design life of 25 years and a total municipal solid waste (MSW) quantity of 2,000,000 tn. Given the size of the landfill, it is assumed that the LFM operations will take place for 10 years and will aim to recover recyclable materials and soil and increase the capacity of the landfill. The MSW content (average, maximum, and minimum concentrations) along with the expected recovery rates through the mining process is given in Table 1.

The excavation of the MSW is made with conventional surface mining equipment, and the haulage of the excavated material is performed using standard dump trucks. The processing unit involves a trommel, a picking line, hand sorting by workers that collect hard and soft plastic, glass, and non-ferrous (primarily aluminum) metals, and a magnet that separates the ferrous metals. In addition, the

Table 1 Composition of waste and recovery rates (in percentages) (Menegaki et al. 2017)

MSW composition	Min value	Mean value	Max value
Ferrous metals (@ 90% recov.)	2.0	4.0	8.0
Non-ferrous metals (@ 85% recov.)	0.3	0.5	0.9
Glass (@ 85% recov.)	2.0	3.5	7.0
Plastics (@ 85% recov.)	3.5	4.1	10.0
Gravel, stones (@ 90% recov.)	5.0	5.0	5.0
Fines, soil (@ 90% recov.)	50.0	50.0	50.0
Residual	33.0	33.0	33.0

Table 2 LFM process technical assumptions (Menegaki et al. 2017)

Description/index	Value	Unit
Hydraulic excavator	1	operating units
Dump trucks	1	operating units
Backhoe loader	1	operating units
Productivity of processing unit	12	tn/hour
Net working hours	6.5	hours/day
Working days (per year)	250	days/year
Productivity/year	19,500	tn/year
Total waste volume	3,300,000	in situ m^3
Specific weight	0.6	tn/m^3
Workforce requirements	13	persons

Table 3 Capital and operating costs for the LFM operations (Menegaki et al. 2017)

Description	Cost (€)
Site preparation and development—Greek typical case	60,000
Administrative costs (per year)	15,000
Capital expenditure for excavation, loading, and hauling equipment	300,000
Capital expenditure of screening and sorting equipment	800,000
Maintenance cost (per year)	22,000
Personnel cost per year (unskilled workers)	14,000
Personnel cost per year (skilled workers)	30,800
Energy cost (diesel fuel, €/lt)	0.95
Energy cost (electric power, €/kWh)	0.09
Water cost (€/m^3)	0.52

Table 4 Estimated benefits from the LFM operations (Menegaki et al. 2017)

	Value (€/tn)		
Benefits	Min value	Mean value	Max value
Ferrous metals	60	80	110
Non-ferrous metals—aluminum	600	700	1000
Non-ferrous metals—copper	1000	1000	2500
Non-ferrous metals—nickel, lead	700	750	1200
Non-ferrous metals (mixed)[a]	660	740	1200
Glass	10	10	15
Plastics (mixed)[b]	100	200	300
Benefit of recovered airspaces	30	30	30
Avoidance of landfill cover material	1.34	1.34	1.34

[a]The analysis is made by taking into account a mix of 75% in aluminum, 15% in nickel and lead, and 10% in copper
[b]The mixed plastics include natural HDPE, mixed HDPE, clear PET, colored PET, etc

processing unit recovers soil that is used as the landfill covered material. The technical assumptions are briefly summarized in Table 2.

Table 3 presents the capital and operating costs of the project. Table 4 provides the prices of the recyclables, along with minimum and maximum estimates, and the economic benefits derived from increasing the landfill disposal capacity and the avoided costs from recovered soil used as the landfill covered material. Finally, the discount rate used is 6%, and the taxation is set to 29%.

Table 5 Projected cash flows

	0	1	2…9	10
Capital costs	1,160,000			
Revenues		624,000	624,000	624,000
Benefit of recovered airspace		351,000	351,000	351,000
Recycling metals, plastics, and glasses				
–Ferrous metals		56,300	56,300	56,300
–Non-ferrous metals		61,400	61,400	61,400
–Glass		5800	5800	5800
–Plastics		136,500	136,500	136,500
Avoidance of landfill cover material		13,000	13,000	13,000
Operating costs		450,000	450,000	450,000
Labor costs				
–Skilled		154,000	154,000	154,000
–Unskilled		112,000	112,000	112,000
Administrative costs		15,000	15,000	15,000
Fuel/energy		107,500	107,500	107,500
Maintenance		58,000	58,000	58,000
Water		3500	3500	3500
EBITDA		169,963	169,963	199,963
Depreciation		113,000	113,000	113,000
Earnings before taxes (EBT)		56,963	56,963	86,963
Taxes (29%)		16,519	16,519	25,219
NOPAT		40,444	40,444	61,744
Cash flow	−1,160,000	153,444	153,444	174,744

3 Results

3.1 *Deterministic and Probabilistic Analysis*

The inflows are about €624,000 per year, and the outflows are €450,000 per year (Table 5). Using a 6% real discount rate, the NPV of the project is estimated at about €1600 and the IRR around 6%; i.e., the project is marginally profitable.

3.2 *ROV Analysis*

The basic inputs of the ROV model using the above-described equations and the results of the DCF analysis are as follows: X = €1.160 million (i.e., the cost of investing in the project), $S0$ = €1.161 million (i.e., the sum of the present values of future cash flows), $\delta t = 1$ year, $T = 4$ years (assuming that the decision will be

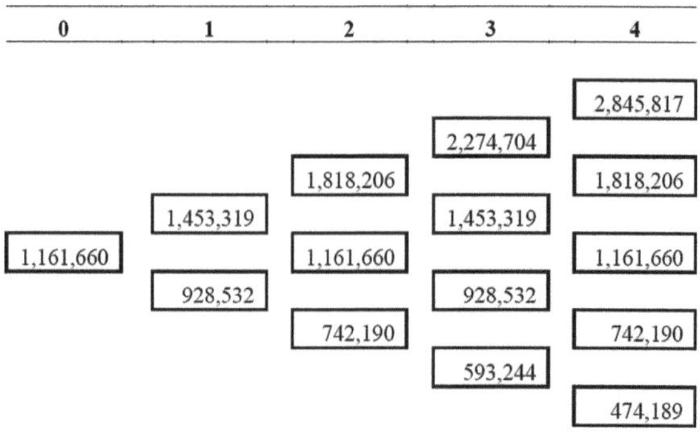

Fig. 2 European option underlying asset lattice of the ROV model

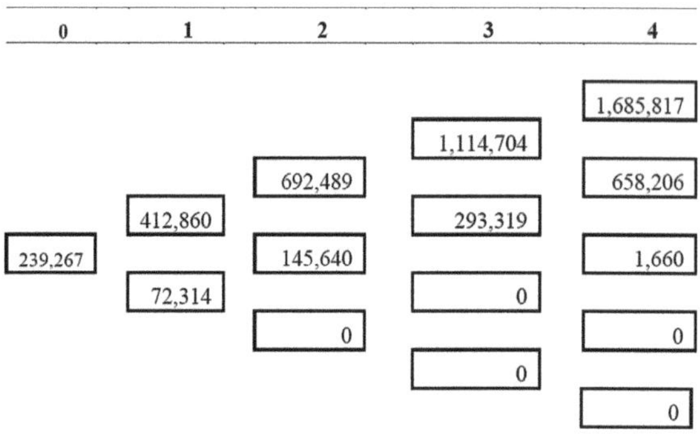

Fig. 3 European option valuation lattice of the ROV model

taken in the next four years), $rf = 3\%$, $\sigma = 22.4\%$, $u = 1.251$, $d = 0.799$, $p = 51.2\%$.

In order to estimate σ, Monte Carlo simulations were conducted considering the most critical technical and economic parameters related to the uncertainty of the estimates (i.e., the price of the recyclable materials and the composition of the MSW). Due to the absence of data about the true distribution of the critical parameters, the triangular distribution was adopted, since there is no requirement that the distribution be symmetrical about the mean.

Starting with the present value of expected future cash flows at year 0 ($S0$), the lattice of the underlying asset is created first by multiplying $S0$ with the up (u) and down (d) factors (Fig. 2), and then, the option valuation lattice is calculated starting from the intermediate nodes back to the starting point (Fig. 3). The Expanded NPV is approximately equal to €240,000. According to the DCF analysis, the static (i.e., deterministic) NPV is €1600. That is, the ROV analysis provides an additional value of €238,400, and it may be considered as the maximum price the buyer would be willing to pay to obtain this option.

4 Discussion and Conclusions

The uncertainty imposed by the particular character of LFM projects (e.g., uncertainty attributed to the composition of the waste and volatility of recycling market) puts certain limits on the DCF analysis. An alternative analytical tool is the Real Options Valuation, which recognizes and values the flexibility in deciding when to proceed with a project based on additional information developed over time.

In the hypothetical LFM project presented, ROV offered a strategic map and revealed the most appropriate decisions to be taken for the project. In this way, it assisted in identifying and capturing an economic value, which is several times higher than the DCF NPV estimates. Actually, the difference between the ROV and the DCF results reflects the value of managerial flexibility. However, it should be remembered that ROV is not "trouble-free." There are certain limitations related to technical and even organizational features of the project. Moreover, problems faced by the "conventional" valuation tools, such as the selection of a risk-free rate, are still present. Thus, ROV should be considered as a supplement rather than a substitute for traditional valuation tools which offers valuable service to decision-makers when there exist high uncertainty and flexibility to influence the timing and or the size of the investment.

Acknowledgements This work was supported by the LIFE + financial instrument of the European Community in the context of LIFE RECLAIM "Landfill mining pilot application for recovery of invaluable metals, materials, land and energy" (www.reclaim.gr), Grant: LIFE12 ENV/GR/000427. The coordinating beneficiary is ENVECO S.A., and the associated beneficiaries are the Municipality of Polygyros, School of Mining & Metal Engineering NTUA, and HELECTOR S.A.

References

Black, F., Scholes, M.: The pricing of options and corporate liabilities. J. Polit. Econ. **81**(3), 637–654 (1973)

Brennan, M.J., Schwartz, E.S.: Evaluating natural resource investments. J. Bus. **58**(2), 135–157 (1985)

Carius, S., Hogland, W., Jilkén, L., Mathiasson, A., Andersson, P.-Å.: A Hidden waste material resource: disposed thermoplastic. In: Proceedings of the 7th International Waste Management and Landfill Symposium, Cagliari, Italy, 4–8 Oct 1999. pp. 229–235

Cossu, R., Hogland, W., Salerni, E.: Landfill mining in Europe and USA. ISWA Year B, Int. Solid Waste Assoc., 107–114 (1996)

Damigos, D., Menegaki, M.: Modern management tools in mining industry in the era of uncertainty. In: 20th International Symposium on Mine Planning and Equipment Selection (MPES 2011), Almaty, Kazakhstan, 12–14 Oct 2011

Damigos, D., Benardos, A., Menegaki, M., Kaliampakos, D., Papagrigoriou, S., Gaitanarou, Z., Stasinos, S.: Assessing the economic viability of landfill mining projects in Greece. In: 4th International Conference on Sustainable Solid Waste Management (CYPRUS 2016), Limassol, Cyprus, June 23–25 2016

Danthurebandara, M., Van Passel, S., Vanderreydt, I., Van Acker, K.: Environmental and economic performance of plasma gasification in enhanced landfill mining. Waste Manag. **45**, 458–467 (2015)

de Neufville, R.: Real options: dealing with uncertainty in systems planning and design. Integr. Assess. **4**(1), 26–34 (2003)

Dimitrakopoulos, R., Sabour, S.: Evaluating mine plans under uncertainty: can the real options make a difference? Resour. Policy **32**(3), 116–125 (2007)

Frändegård, P., Krook, J., Svensson, N., Eklund, M.: A novel approach for environmental evaluation of landfill mining. J. Clean. Prod. **55**, 24–34 (2013)

Frändegård, P., Krook, J., Svensson, N.: Integrating remediation and resource recovery: On the economic conditions of landfill mining. Waste Manag. **42**, 137–147 (2015)

Hogland, W., Marques, M., Thörneby, L.: Landfill mining—space saving, material recovery and energy use. In: Proceedings of Seminar on Waste Management and the Environment-Establishment of Cooperation Between Nordic Countries and Countries in the Baltic Sea Region, Kalmar University, Kalmar, Sweden, 5–7 Nov 1997. pp. 339–355

Hogland, W., Marques, M., Nimmermark, S.: Landfill mining and waste characterization: a strategy for remediation of contaminated areas. J. Mater. Cycles Waste Manag. **6**, 119–124 (2004)

Jones, P.T., Geysen, D., Tielemans, Y., Passel, S.V., Pontikes, Y., Blanpain, B., Hoekstra, N.: Enhanced landfill mining in view of multiple resource recovery: a critical review. J. Clean. Prod. **55**, 45–55 (2013)

Kapur, A., Graedel, T.E.: Copper mines above and below ground. Estimating the stocks of materials in ore, products, and disposal sites opens up new ways to recycle and reuse valuable resources. Environ. Sci. Technol. **40**, 3135–3141 (2006)

Kelly, S.: A binomial lattice approach for valuing a mining property IPO. Q. Rev. Econ. Finance **38**, 693–709 (1998)

Kodukula, P., Papudesu, P.: Project Valuation Using Real Options: A Practitioner's Guide. J. Ross Publishing, N. Andrews Way, Florida (2006)

Krook, J., Svensson, N., Eklund, M.: Landfill mining: a critical review of two decades of research. Waste Manag. **32**, 513–520 (2012)

Kurian, J., Esakku, S., Nagendran, R.: Mining compost from dumpsites and bioreactor landfills. Int. J. Environ. Technol. Manag. **7**, 317–325 (2007)

Lee, G.F., Jones, R.A.: Use of landfill mining in solid waste management. In: Water Quality Management of Landfills Conference. Water Pollution Control Federation, Chicago, IL (1990). p. 9

Menegaki, M., Benardos, A., Kaliampakos, D., Tsakalakis, K.: Is landfill mining a new prospect for the mining sector? In: 24th International Mining Congress and Exhibition of Turkey, Antalya, Turkey, 14–17 Apr 2015, pp. 101–111

Menegaki, M., Benardos, A., Damigos, D., Kaliampakos, D., Tsakalakis, K.: Emphasizing the role of e-waste in the financial profitability of landfill mining projects. MATTER Int. J. Sci. Technol. **3**(2), 122–144 (2017)

Merton, R.C.: The theory of rational option pricing. Bell J. Econ. Manag. Sci. **4**, 141–183 (1973)

Moyen, N., Slade, M., Uppal, R.: Valuing risk and flexibility: a comparison of methods. Resour. Policy **22**(1/2), 63–74 (1996)

Mun, J.: Real options analysis—tools and techniques for valuing strategic investments and decisions, 2nd edn. Wiley, Hoboken, NJ (2006)

Obermeier, T., Hensel, J., Saure, T.: Landfill mining: energy recovery from combustible fractions. In: Proceedings Sardinia '97, Sixth International Landfill Symposium, Cagliari, Italy (1997). pp. 569–578

Poulimenou, N., Damigos, D.: Corporate environmental performance and economic crisis. In: Fifth International Conference on Environmental Management, Engineering, Planning & Economics (CEMEPE 2015), Mykonos Island, Greece, 14–18 June 2015, pp 566–572

Sabour, S.A., Dimitrakopoulos, R.G., Kumral, M.: Mine design selection under uncertainty. Mining Technol. IMM Trans. Sect. A **7**(2), 53–64 (2008)

Samis, M., Davis, G., Laughton, D., Poulin, R.: Valuing uncertain asset cash flows when there are no options: a real options approach. Resour. Policy **30**(4), 285–298 (2006)
Slade, M.: Valuing managerial flexibility: an application of real-option theory to mining investments. J. Environ. Econ. Manag. **41**(2), 19–233 (2001)
Zhou, C., Gong, Z., Hu, J., Cao, A., Liang, H.: A cost-benefit analysis of landfill mining and material recycling in China. Waste Manag. **35**, 191–198 (2015)

MIX
Papier aus verantwortungsvollen Quellen
Paper from responsible sources
FSC® C105338

If you have any concerns about our products,
you can contact us on
ProductSafety@springernature.com

In case Publisher is established outside the EU,
the EU authorized representative is:
Springer Nature Customer Service Center GmbH
Europaplatz 3, 69115 Heidelberg, Germany

Printed by Libri Plureos GmbH
in Hamburg, Germany